KNOX

People ◐ Planting ◑ Programs

Good Food, Strong Communities

Good Food, Strong Communities

Promoting Social Justice through Local and Regional Food Systems

STEVE VENTURA AND
MARTIN BAILKEY, EDITORS

University of Iowa Press, Iowa City

University of Iowa Press, Iowa City 52242
Copyright © 2017 by the University of Iowa Press
www.uipress.uiowa.edu
Printed in the United States of America

Design by Ashley Muehlbauer

The University of Iowa Press is a member of Green Press
Initiative and is committed to preserving natural resources.

Printed on acid-free paper

Library of Congress Cataloging-in-Publication Data
Names: Ventura, Steve, 1955– editor. | Bailkey, Martin, editor.
Title: Good food, strong communities : promoting social
justice through local and regional food systems / Steve
Ventura and Martin Bailkey, editors.
Description: Iowa City : University of Iowa Press, [2017] |
Includes bibliographical references and index.
Identifiers: LCCN 2017005545 | ISBN 978-1-60938-543-9 (pbk)
| ISBN 978-1-60938-544-6 (ebk)
Subjects: LCSH: Food supply—United States. | Local foods—
United States. | Urban agriculture—United States.
Classification: LCC HD9005 .G66 2017 | DDC 338.1/973—dc23
LC record available at https://lccn.loc.gov/2017005545

To Jerry Kaufman (1933–2013), who inspired so many.

You inspire us still, with your wisdom and integrity.
We are grateful to have begun this journey with you—our dear colleague,
mentor, and friend.

Professor Jerry Kaufman taught in the Department of Urban and Regional Planning at the University of Wisconsin–Madison, pioneered the field of food system planning, and was the board president of Growing Power.

Contents

CHAPTER 3. GROWING URBAN FOOD

CHAPTER 4. DISTRIBUTION: SUPPLYING

Foreword

Urban agriculture has grown tremendously in the past five years. From the increase in the number of folks who want to grow food in their backyards to the industry's recognition by corporations and the funding for growing systems from food wholesalers, we have seen so many positive changes. People feel able to make a living from this type of farming because we've shown that they can scale it up.

We need increased local food production through urban agriculture in order to provide enough good food for our communities. There is a reason that Growing Power, a Milwaukee-based urban farming project, scaled up production. I wanted to prove that it is possible to change the dynamics of a community by growing enough food. I wanted to prove that you could put people to work and be part of local economic development. I wanted to create an industry that would help communities. We don't have enough farmers right now to grow enough good food through sustainable production in this country. That's what we have to do.

In this project we looked at the current state of urban agriculture and food production by working with organizations from around the country to understand the challenges they face. How much commonality is there across the groups? It was important to ask that question in a context that allowed us to quantify the answers in order to create models that are best practices, not only for the seven cities we studied but also for communities across the country.

There are a lot of commonalities in experiences among these groups. Soil is the main problem in urban agriculture. A high percentage of the farmable space is asphalt and concrete. You can't dig into it. One of the things the groups really struggle with is getting high-quality compost to grow in. Other parts of the country don't have a lot of open land like what we have in the Midwest, and the land they do have is very expensive. Another common challenge is the lack of funding available to scale up production, particularly for entrepreneurial farmers.

One of the strengths of the small-scale organizations we partnered with is that they can control their operations as they increase production, rather than beginning with a large-scale organization that is missing important pieces. Building infrastructure is not necessarily the answer; you need farmers with experience.

There is a lack of farmers who can work in multicultural environments or with communities of color. Farmers in these communities need the appropriate technical knowledge and social skills to succeed. We have to find people who have absolute passion. No matter what happens, they'll need to stick to it long enough to become successful, because without that deep, deep passion, farming is a tough thing. If you don't have that passion, to really want to do it and see it through, you probably won't do it for very long. I see a lot of organizations that give up and scale down instead of scaling up.

By examining the best practices from our partnering organizations, we can identify opportunities for academic institutions to support communities as they work to improve their local food systems. This project was a good start to building relationships between the university and community-based organizations. Low-income communities are looking for help to plan their projects, get them running, and build infrastructure for their organizations. Universities have some skill sets that the community doesn't have but desperately needs.

There is a lot of mistrust in those communities, though, because there is a history of dysfunctional relationships between academic institutions and community-based organizations. Over the years, universities haven't taken the time to understand those communities before stepping into them. Each of those communities is a little different, and so is the leadership in those communities. Partnerships must be led by the people who have been doing the work there, who have built relationships. If a university turns to the largest organization in the community, it might not be connected to the folks there. There are not a lot of people of color in the leadership of many organizations around the country, so there are people who operate organizations in communities of color but are not part of the community.

Partnering with an organization because a grant opportunity pops up may lead to exploitative relationships if the group does not have the strength and capacity to ensure that its members receive equal access to grant resources. Some grants are so lopsided that by the time funding is distributed through the university structure, only a little chunk is left at the end for the community. The organization often does the heavy lifting in the work that universities want to study. We have to share the funding in equitable ways, and a lot of coordination must come from the communities.

Throughout this project, because we focused on food justice in food systems, we wanted to make sure that we had some food justice in the overall approach. We had to push back so that meetings would begin with the University of Wisconsin listening to the community groups and then bringing them into the conversation. One of the main messages is that universities can't go into communities until they're invited. We had some hard lessons at first, but I am proud that the university learned a lot about how to work in communities during the project.

It's something that will take time. Sometimes people think that things happen because you've written a proposal and secured a grant, when often the end of that grant is just the beginning of the actions that can take the work a lot deeper. We need more projects like this in which we practice the art of building relationships into partnerships. Now people know each other; this is the start of something that could be much bigger.

We're just in the infancy stages of building a new food system. This book can be used to do follow-up work and to serve as a tool for universities, community-based organizations, city governments, and students at any level to understand the challenges of urban agriculture and continue best practices. If the book serves as an educational text, that's a very powerful thing to do. If it can help to build an equitable, just, productive, and healthy food system, then we've done our job.

—Will Allen
CEO, Growing Power

Good Food, Strong Communities

Preface

Writings about food systems in the United States have proliferated in the past decade. Much of this work has decried the consequences of a system that is oriented to producing massive amounts of food as cheaply as possible. Some pundits even put *food* in quotes, questioning whether the industrial food system that produces the vast majority of food in this country meets people's needs from the perspectives of nutrition, health, environment, and culture. Of particular concern is the belief and observation that the negative consequences of mainstream food production and marketing, particularly the lack of easy and dependable access to fresh and healthy food, fall more on the poor and on people of color. While recognizing some of the problems of mainstream food systems, this book is more about the latter premise: that current food systems are another manifestation of the racism and economic disadvantage suffered by communities of color, and that changing these systems will contribute to building a more positive social structure.

Our book is an outgrowth of the Community and Regional Food Systems (CRFS) project, which was initiated in 2011 in response to a US Department of Agriculture (USDA) call for proposals around the issue of food security. The "we" who provide the majority of writing in this book come from academic organizations, primarily the University of Wisconsin–Madison (UW–Madison) and the University of Wisconsin Extension (UWEX). However, the project included scores of community organizations as partners. Growing Power and the Michael Fields Agricultural Institute were major partners. Key participants in seven cities—Madison, Milwaukee, Boston, Cedar Rapids (Iowa), Chicago, Detroit, and Los Angeles—helped us connect with community organizations to learn about and foster food system innovations and successes. We also helped support training and outreach through a Minnesota organization that has developed several community-based food projects in Minneapolis.

Some of these organizations were involved explicitly through our community engagement projects and Innovation Fund subcontracts; many others willingly shared their ideas and experiences, and we are grateful to all of them. We include their words when feasible and hope that we are effectively channeling their voices otherwise. An unstated but cogent goal of the CRFS project was to learn how to work together: how to recognize and build on each other's strengths to create partnerships. In the words of academics, this project was participatory and community-based research and outreach; in the words of Growing Power farmer-in-chief Will Allen, one of the project's principal investigators, it was steps toward the creation of an "urban extension."

We hope that something in this compilation will appeal to or be useful for community-based practitioners, food system activists, extension service educators, academics, food policy advocates, and anyone who understands and appreciates the challenges involved in creating just, healthful, and sustainable food systems.

A BRIEF OVERVIEW OF THE CRFS PROJECT STRUCTURE AND APPROACHES

CRFS was proposed to the USDA as an integrated project that would include research, outreach, and education. We explicitly included a fourth aspect in our proposal, advocacy, recognizing that meaningful change in food systems was unlikely to occur without it. We assembled a team with expertise in a wide range of disciplines and backgrounds as well as experience in all four aspects. The project was led by co–principal investigators Steve Ventura of UW–Madison and Will Allen of Growing Power. Day-to-day guidance was provided by project comanagers Greg Lawless of UWEX and Martin Bailkey of Growing Power. Our core team of researchers, educators, and advocates (which included many of the authors in this book) evolved over time and always involved representatives of these three organizations, plus Margaret Krome of the Michael Fields Agricultural Institute.

At the risk of raising eyebrows among our benefactors at the USDA, we can now say that two components of the research we originally proposed were actually always nonstarters. First, we proposed to do a systems analysis of community and regional food systems. This is a challenge akin to the "blind men and the elephant" description problem (everyone focuses on a different aspect), with the added complication that the elephant is galloping rather than standing still. A growing local and regional food movement, changes in federal policy, place-based food justice and food sovereignty efforts, and other influences mean that US food systems are rapidly evolving. The framework for understanding community food systems, described in chapter 1, starts with systems analysis: breaking down the food systems into their components. We depict the elements of a food supply chain and the values that drive it, so in a graphic sense we have accomplished the objective. This framework provides a way to talk about the dynamic influences on and changes in the food systems—not the detailed descriptions of processes, methods, and products envisioned by systems analysts, because these are contextually different and dynamic.

Our second nonstarter was case-study research. Although our proposal was based on participatory research methods, it was apparent even before we officially began that our community partners did not want to be studied. As described in chapter 13, they wanted interaction on an equal basis, and they expected that the researchers would come into their communities on each community's terms. This requirement helped refine one objective of our original proposal, the Innovation Fund projects, and led to the creation of another type of research and outreach, our community engagement projects.

The Innovation Fund projects were minigrants to ten organizations to, in essence, do good food system work. These organizations were given $50,000 each, with few strings attached. We asked them to construct projects that they thought would advance their work but that might not be possible without additional funding. Our interest was in providing the freedom to innovate, and all we asked was the right to observe and distill good ideas from what they accomplished. Aspects of these projects are noted in chapters and sections throughout this book. The following list summarizes the projects by city, project name, organization, and project director:

- Boston: Dudley Grows: Increasing the Availability of Produce in Retail Settings in the Dudley Street Neighborhood (The Food Project, Sutton Kiplinger)
- Cedar Rapids: Evaluating the Potential of Neighborhood Assets to Meet the Food Needs of Cedar Rapids Neighborhoods (Iowa Valley Resource Conservation & Development, Jason Grimm)
- Chicago: Community-Based Assistance to Spur Food Innovation Activity (Chicago Food Policy Action Council, Erika Allen)

- Detroit: Ujamaa Food Co-op Development (Detroit Black Community Food Security Network, Malik Yakini)
- Los Angeles: The Sankofa Project (Community Services Unlimited Inc., Neelam Sharma)
- Madison: The Madison Community Orchard Project (Community GroundWorks, Karen von Heune)
- Milwaukee: Urban Agriculture and Neighborhood Food Distribution (Walnut Way Conservation Corporation, Sharon Adams)
- Milwaukee: Economic Evaluation of Urban Farming (Milwaukee County Extension, Ryan Schone)
- Milwaukee: Local Food Policy Assessment for the City of Milwaukee (Center for Resilient Cities, Marcia Caton Campbell)
- North Branch, MN: The North Circle Project (Women's Environmental Institute, Jacquelyn Zita)

The community engagement projects were community-driven activities that involved academic partners in some aspect. Instead of going into a community and saying, "We'd like to study you," the members of our core team went to meetings and talked to community leaders to learn what issues they thought were important. This led to discussions of what kinds of resources and expertise CRFS partners could contribute, and in more than thirty cases there was collaboration on a wide range of projects and activities. The assessment of Detroit's Uprooting Racism, Planting Justice group in chapter 9 is a good example, as is the section of chapter 14 that describes a coalition's activities to improve the products available in supermarkets serving a Latino community in south Milwaukee. In Los Angeles a CRFS project team helped the Los Angeles Food Policy Council create an evaluation protocol for its groundbreaking Good Food Purchasing Program, as described in chapter 12. The community engagement projects did not have a defined methodology. Their characteristics evolved from the participants and the circumstances, and we learned by observing the discussions and outcomes.

The main educational initiative of CRFS is described in chapter 11. We created an urban agriculture curriculum for UW–Madison's very successful Pre-College Enrichment Opportunity Program for Learning Excellence (PEOPLE), a pre-college pipeline for students of color and low-income students. The project also supported the education of many college students, with an effort to support students of color (five of the ten graduate students and twelve of the twenty-eight undergraduates employed on the project). We believe that this has led to an enduring legacy of the project—a pipeline starting in high school that helps minorities recognize career paths in food and agricultural systems and supports their progress through higher education.

Outreach and training came through two pathways in the CRFS project. The first was support for Growing Power's long-standing training programs: From the Ground Up workshops on community food systems and Commercial Urban Agriculture farmer training, including new lessons taught by university faculty and students. The second was the development of two new Internet resources: the *Urban Agriculture Manual* (UWEX 2017) and the *Community Food Systems Toolkit* (UWEX n.d.).

The advocacy aspects of our project also had two main thrusts. We worked on policies supportive of urban agriculture at all levels, from local to federal, particularly under the leadership of the Michael Fields Agricultural Institute. We also helped build and bolster institutions oriented to sustainable urban food systems, such as the Cooperative Institute for Urban Agriculture and Nutrition in Milwaukee (see chapters 10 and 14) and the eXtension Community, Local, and Regional Food Systems Network.

When asked to describe what CRFS was about and how it worked, the project team members often explained, "It's a big, sprawling project with many moving parts." What we did could be described as participatory action research, as community-based research, and by various other academic terms. But specific methods evolved to meet changing circumstances; keeping in mind the ultimate purpose of addressing food security, we found ways to make progress and share what we learned.

ADDRESSING FOOD SECURITY THROUGH COMMUNITY FOOD SYSTEMS

The goal of this book is to share ideas and stories about community efforts to improve food security in large urban areas of the United States through community food systems. According to Hamm and Bellows (2002, 35), community food security is a situation "in which all community residents obtain a safe, culturally acceptable, nutritionally adequate diet through a sustainable food system that maximizes community self-reliance and social justice."

How community and regional food systems can address food security is the theme of the first chapter of this book as well as the goal of the CRFS project.

Chapter 1 defines key concepts and introduces a framework of community food systems, thus providing the language and structure for subsequent chapters. The term *food supply chain* is often used to describe all the activities that go into creating and consuming food—*from farm to fork*, in colloquial terms. As we further explored local and regional food systems, however, it was apparent that food supply chains operate in a broader context. Also, the chains we found in communities were driven by multiple values. Chapter 1 introduces a graphic representation of food supply chains, which has proven to be a useful framework to understand and communicate what's going on in community and regional food systems.

The elements of the framework—the food supply chain, the broader context in which it operates, and the values that drive systems—are the basis for a detailed exploration of food system innovations in chapters 2 through 8. Various communities—neighborhoods, municipalities, school systems, entrepreneurs, and others—seek to create sustainable and just food systems that provide healthy food. These efforts address one or more aspects of food production, processing, marketing, and consumption. In urban areas, efforts to create sustainable and just community food systems often go beyond supply chain components by addressing the social, environmental, and economic issues underlying food insecurity. The framework provides a structure to delve deeply into examples that describe the process and products of innovations along with the associated values, as explored in chapters 9 and 10.

The last chapters of the book distill some of the lessons learned from our CRFS project. These discussions include the role of education and policy in food system change (chapters 11 and 12) and highlight the incredible and wonderful diversity of efforts and approaches. When reflecting on the project activities in seven cities, we also saw numerous overarching themes that run within and among the cities. Although these themes are found in the discourse and activities of each city, particular locations provide exemplary cases, such as the experiences of collaborators in Detroit (chapter 9) and Chicago (chapter 13). This experience has been enriched by the willingness and interest of people in these communities to work with university partners. An important lesson is that academic-community partnerships can lead to mutual benefits for all involved: researchers, educators, and practitioners. We hope that this will ultimately lead to better food systems for those most in need: the consumers in underserved areas of our major cities.

CHAPTER ONE

Connections between Community Food Security and Food System Change

STEVE VENTURA AND MARTIN BAILKEY

In this chapter CRFS Project Codirector Steve Ventura and Project Comanager Martin Bailkey define community food security, introduce the key concepts of innovation and collective impact, acknowledge the guiding importance of social justice to our work, and present a systems framework for communicating the essential elements of food supply chains and associated values in community and regional food systems.

Near the beginning of the Community and Regional Food Systems project, we participated in the Growing Power Urban and Small Farm Conference in Milwaukee. We videotaped some participants talking about their personal definitions of key concepts such as food security. A youth organizer for a community nonprofit group in Oakland, California, said, "Community food security is really a food system in which the folks who are eating the food know the people who are making the food and know where to get good food. The food is grown in their communities, and the money that is used to purchase the food goes back to nourish the community."

This simple but elegant definition of community food security comes from an organizer who works in a poor community that struggles with the consequences of a national food system that is indifferent to its needs. The paradox apparent to anyone looking closely at these struggles is that the same people who may at times go hungry also suffer the consequences of an excess of poor-quality food, such as obesity, diabetes, and heart disease; the list of health problems associated with poor nutrition is extensive and frightening.

The invidious aspect of this paradox is the lack of choice. Although most people in modern America have access to ample food, this is not the case for everyone. Another food system activist at the Growing Power conference said, "Food security is a basic human right." He and many others consider the effort to develop food-secure communities as akin to other efforts to secure human rights. Related terms such as *food justice* and *food sovereignty* are also used by activists, connoting the strong link of food system challenges to race and poverty in the United States. The community food systems they seek to create support human rights, including healthiness, dignity, and livelihood.

We started the CRFS project in 2011 by adapting an international definition of food security to reflect community-oriented goals. The Food and Agriculture Organization (FAO) has an extensive discussion of terminology related to food and nutrition security (Committee on World Food Security 2012). It concludes, "Food and nutrition security exists when all people at all times have physical, social, and economic access to food, which is safe and consumed in sufficient quantity and quality to meet their dietary needs and food preferences, and is supported by an environment of adequate sanitation, health services, and care, allowing for a healthy and active life."

The FAO uses the terms *adequate, available,* and *accessible* as shorthand for this definition. We extend this definition to what we call the five As of food security, specifically aimed at a community level:

- **Available.** Community and regional production, processing, and distribution.
- **Affordable.** Reasonable real and perceived costs to the populations in need.
- **Accessible.** Food for people in stores, restaurants, markets, shelters, and pantries.
- **Appropriate.** Nutritious, safe, culturally relevant, appealing food.
- **Acceptable.** Sustainable environmental, social, and economic effects of all components.

The US Department of Agriculture uses a definition of food security geared to the household level. It ranges from "high food security: no reported indications of food-access problems or limitations" to "very low food security: reports of multiple indications of disrupted eating patterns and reduced food intake." The USDA definitions of food security and insecurity include a reference to "nutritionally adequate and safe" food and a requirement that the means of acquiring food is socially acceptable. These definitions are primarily oriented to the *amount* of food that is available in a household, which is made apparent by the description of how food insecurity is measured: "Food insecurity—the condition assessed in the food security survey and represented in USDA food security reports—is a household-level economic and social condition of limited or uncertain access to adequate food" (Coleman-Jensen 2006).

Although the USDA definition of food security falls short of all the community and public health values that we believe should be part of food security, it is used by several authors in this book because it is the prevailing indicator used by emergency food providers and others involved in addressing food system issues in underserved communities. As noted in the preface, we also frequently use the definition provided by Michael Hamm and Anne Bellows in 2002, since it combines community and household, food quality and quantity, and food justice and sovereignty.

The activities, ideas, and passions expressed in this book are aimed at understanding the extent to which community-based food systems improve one or more aspects of food security. This includes asking the following: In what ways do activities in the food system address the root causes of food insecurity? This question means that we must look at community food *systems* as a whole. This is challenging, because food systems are dynamic, diverse, and difficult to understand in their entirety. They can be viewed from many perspectives, including the biophysical aspects of the food supply chain and the broader environments in which the system operates: economic, legal, political, and social.

Figure 1 expresses this complexity. At the national kick-off meeting of the CRFS project in 2012, we asked representatives of each of the seven participating cities to draw diagrams of their local food systems. Food system activists from Detroit did a great job expressing key issues and relations in their city, including the broad interaction of local economies, sustainability, and issues of race and power.

It's no accident that social justice is drawn as the central issue in Detroit. Activists used the term *food sovereignty* for "rebuilding a new food system with new values." For them, this includes building a new economy around food to alter "a history of class struggle" and developing innovative ways to access and use the tens of thousands of acres of abandoned and vacant property within the city. In the figure, the terms with squiggly arrows outside the food system circles are accurate reflections of the food system challenges we've seen across the country:

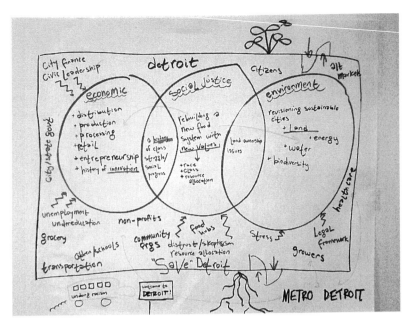

FIGURE 1. Depiction of Detroit food system developed by Detroit participants at CRFS project national meeting, Milwaukee, May 2012. (Courtesy of Greg Lawless, CRFS)

leadership, finances, unemployment, inadequate education, the legal framework (policy), transportation, and environmental stress such as land contamination. Though not unique to Detroit, distrust and skepticism of outside assistance—people who want to "save" the city—is also a challenge.

FOOD JUSTICE AND FOOD SOVEREIGNTY

The previous section mentioned two concepts—food justice and food sovereignty—that are key to this book and merit more explanation. Direct engagement in a community-based food system is a political action in the sense that an individual or a household's ability to access and consume nutritious food correlates with its economic status and thus its power within society. The politics of food plays out in many ways, and it affects not just the realm of consumption (e.g., federal government support of hunger programs) but also that of distribution and production, such as the effects of federal immigration policies on farmworkers.

Explicitly and implicitly, the concept of community food security contains the related concept of *food justice*. Based on ideas from established movements in civil rights, environmental justice, community empowerment, and workers' rights, among others, food justice has been defined by Gottlieb and Joshi (2010, 6) as "ensuring that the benefits and risks of where, what, and how food is grown and produced, transported and distributed, and accessed and eaten are shared fairly."

The Institute for Agriculture and Trade Policy (IATP 2012) offers another definition of food justice as "the right of communities everywhere to produce, process, distribute, access, and eat good food regardless of race, class, gender, ethnicity, citizenship, ability, religion, or community."

The IATP also specifies within its definition the freedom from exploitation, fair labor practices toward food system workers, and the recognition of underlying values (respect, empathy, pluralism), racial justice, and gender equity. The key point of the IATP definition, however, is the idea of equitable participation in the food system as a basic human right. This recalls the expression of food as a basic human right in the United Nations Universal Declaration of Human Rights of 1948, which set a path for later concepts of food security and, more recently, food sovereignty (Chen, Clayton, and Palmer 2015).

As with food justice, a second IATP (2012) definition also posits the related concept of *food sovereignty* as a right of all people: "Food sovereignty is the right of people to define their own food, agriculture, livestock, and fisheries systems."

The concept of food sovereignty can be linked to that of community self-determination—specifically, the opportunity for any community, particularly one with few resources, to free itself from outside control over how and what it eats every day. Though not always acknowledged as the conceptual foundation, food sovereignty underlies almost all grassroots efforts to initiate local food projects, particularly when they are understood and publicly announced as alternatives to conventional, corporate food options that are of poor nutritional quality and that extract money from communities.

With its roots in peasant and worker movements of the global South, food sovereignty is linked to community food security through the shared political dimensions of these groups, and thus by necessity it involves social activism. Social empowerment is achieved by organizing time, resources, and passions around food production, access, and consumption. The breadth of alternative food system activities results in a similar range of food sovereignty examples—from tiny "guerrilla gardens" on vacant urban land to La Via Campesina, the global movement of peasants, migrants, landless people, and small- to medium-scale farmers operating through more than 160 organizations in seventy countries. Food sovereignty activists, to a large degree, draw from the food cultures and traditions of the global South as models for system change in developed parts of the world (Roberts 2008). In addition, food sovereignty actions can reveal and address the power differentials exhibited through the everyday presence of discrimination, racism, and environmental injustice in the global food system.

Though sometimes overlooked in the focus on food quality and access for all, the rights of food system workers bear special consideration as a component of food justice, if only for the sheer number of workers involved. A Food Chain Workers Alliance (2012) report stated that roughly 16 percent of the US work force, almost twenty million workers, are employed in five areas of the food system: production, processing, distribution, retail, and service. These jobs are characterized by poor wages and working conditions and by limitations to advancement linked to discrimination by race or immigrant status.

Dismantling Racism through an Alternative Food System

The CRFS project, in its search for innovative food system practices, considered the linked concepts of community food security, food justice, and food sovereignty as lenses with which to assess examples of food system innovation in the United States, a nation that typically takes food access for granted. The contributions of the project's community partners, many of them people of color working in communities of color, provided important guideposts. The work of Growing Power, a primary CRFS organizational partner, has played an important role. Growing Power, founded by African American farmer Will Allen in 1993 and managed by a multicultural staff, has dedicated itself to identifying and addressing the pervasive influence of structural racism on the food system. This commitment was first formally adopted by the organization through strategic planning in 2006, making explicit its approach toward inclusivity and social justice that Growing Power had displayed from the start.

An antiracism perspective underlies all of Growing Power's work. The Growing Food and Justice for All Initiative, a program of Growing Power, introduced members of the CRFS project team, many of whom were white, to what Growing Power experienced through working in communities of color—the same communities in which the CRFS project sought examples of food system innovation. Thus, one role assumed by Growing Power and the Growing Food and Justice for All Initiative as a CRFS project partner was to figuratively and practically bridge the gap between people of color and the prevailing whiteness of the alternative food movement in the United States (Guthman 2008, 2011; Slocum 2006; Morales 2011a), particularly within academia. To see how this process transpired in Chicago, see chapter 13, which features contributions from Growing Power and the Chicago Food Policy Action Council.

Dismantling the system of interlinked societal and economic practices that forms structural racism suggests an incremental deconstruction of its components, the institutions that maintain the system: education, employment, the media, the transportation infrastructure, the justice system, and corporations, among others. The concept of community food systems represents an arena in which those victimized by racist actions that are euphemistically labeled as something else (e.g., the closing of corporate-owned inner-city supermarkets because of their limited building footprints or the "low buying power" of urban neighborhoods) can incrementally begin to reverse the established patterns of power and assume greater control over their lives. This marks a significant shift in perspective and approach from the top-down, antipoverty and antihunger government programs of the 1960s and later.

The primacy of food as a daily element of life—and the traditions and cultural practices around growing, preparing, and consuming it—means that community-based food system activism is a viable, proactive strategy for combatting the effects of food injustice and outright racism. (Chapter 9 describes a case study in Detroit.) In her study of the African American farmers of D-Town Farm in Detroit, Monica White (2010) of the University of Wisconsin–Madison (a CRFS project affiliate) concluded, "Ending a relationship that is dependent upon the whim of a supermarket chain or a politician's popularity, [D-Town] farmers *have decided to control their own food supply and their own movement* . . . they choose to provide food for themselves and their community. In providing an alternative behavioral option to dependence on the state, *they prefer to act in ways that demonstrate agency and empowerment*" (emphases added).

Another indicator of the importance of a food justice foundation to community food systems is the degree to which food justice issues serve as an entry point to social activism for youth and young adults. Several community-based organization partners in the CRFS project engage youth as a significant part of their missions. For example, the cover photograph of *Food Justice* by Gottlieb and Joshi (2010) shows thirteen confident, self-assured members of The Food Project's Summer Youth Program (now called the Seed Crew). Effective food activism entails understanding the nature of the current food system; thus youth gain practical experience and valuable insight into how the world works through helping to change the food system.

The five-year term of the CRFS project meant that the project staff had no realistic expectation of achieving a full understanding of how structural racism in the United States influences the food system. That remains an ongoing effort. To the best of their abilities, however, the contributors to this book tried to frame their accounts through lenses informed by the antiracism approaches of Growing Power, the Growing Food and Justice for All Initiative, and CRFS partner organizations across the nation.

The Sankofa Project, Community Services Unlimited, Los Angeles

NEELAM SHARMA

The Sankofa Project of Community Services Unlimited (CSU) is one of the ten Innovation Fund projects directly supported by the CRFS project. Founded by the Southern California chapter of the Black Panther Party in 1977, CSU has a radically different history from that of the vast majority of the other 1.2 million nonprofit organizations in the United States. CSU was created to help the Panthers build a revolution—in other words, to help oppressed people come together to transform their conditions. After the demise of the Panthers, CSU was kept intact as a vehicle to keep "serving the people" of South Los Angeles through volunteer community-supported efforts. As a tax-exempt nonprofit organization, CSU is structurally part of the nonprofit industrial complex, described by author and journalist Mumia Abu-Jamal on the book jacket of *The Revolution Will Not Be Funded* as "an unseen web of money and power that tries to undermine people's struggles."

Indeed, over the last fifteen years CSU has successfully received foundation and government grants. It is this potentially empowering contradiction that the Sankofa Project seeks to investigate, illuminate, and learn from to help us keep our eyes on the prize and not get caught in the dead end of chasing money and becoming co-opted. *Sankofa*, in the Twi language of Ghana, means "reach back and get it"; in other words, we look back to our past to understand our present conditions, then map out a course for the future. The Sankofa Project is an ongoing, participatory, multimedia, and multidisciplinary project designed to do just this. It demonstrates that the current generation of food justice work is not new but follows a long tradition of grassroots resilience and creativity. Together with our From the Ground Up (FGU) training program, Sankofa connects the young people of South Los Angeles to this lineage, wherein they see their ancestors, their communities, and themselves. FGU youth are integral to the project and are actively involved in multiple tasks, including organizing video footage and other materials, conducting research, and creating and implementing surveys.

The project officially launched in October 2015 at the William Grant Still Arts Center with an exhibition of archival photographs and other materials spanning CSU's history, along with videos of our current activities. Although more work remains, the Sankofa Project has helped us gain and document important insights on how our history affects the way we understand and choose to work in the world around us. For example, the way we hire and train staff is unlike that of most nonprofit groups, because we de-emphasize qualifications and experience and emphasize understanding the bigger social justice issues we are grappling with (an approach for which we have been criticized by certain funders).

The Sankofa Project's comparative research, through site visits and interviews, of two sister organizations—the Detroit Black Community Food Security Network and the Social Justice Learning Institute in Inglewood, California—revealed something interesting about our own hiring practices. Both organizations share the objectives of CSU, with histories deeply connecting them to past struggles and the Black Panthers. Our three organizations have similar hiring practices, rooted in a deep commitment to building skills and jobs among neighborhood residents, and a general mistrust of the highly professionalized nature of the nonprofit industry. This important revelation helped all three organizations understand that our hiring strategies were not quirky practices that could be easily dismissed.

In creating our own research about ourselves through the CRFS project, we are building and deepening connections to help form our vision of equitable, healthy, and sustainable communities that are self-reliant and interrelating and in which everyone has the support and resources they need to develop to their fullest capacity.

The negative consequences of our current food systems for individuals and communities are a manifestation of broader and deeper societal forces associated with poverty and racial discrimination. Changing the food system does not directly address these underlying forces, but it can alleviate some of the symptoms and provide a pathway for progress. One way to address food security and food justice challenges is to change food systems: rebuild connections between people and agriculture, rejuvenate curricula so children know about food, rethink supply chain logistics, reorient corporations to serve people, reform laws that now create incentives for the dysfunctional aspects of food systems, and renew respect for the central role that food plays in society.

Change in our food systems will happen in myriad ways, through on-the-ground innovations and new practices, new ways of processing and delivering food, changes in public policies and programs, changes in institutional behaviors, and many other ways. In this book we have adopted the term *collective impact* for change that occurs from many people and organizations working on many aspects of a multifaceted issue.

Collective Impact

Collective impact is defined as "the commitment of a group of important actors from different sectors to a common agenda for solving a specific social problem" (Kania and Kramer 2011, 36). The group members may have a shared location, but more important, they have common interests and, in the case of food security, a common identification of the issues described in this chapter and a shared belief in the need for food system change. Community-driven change efforts are typically asset based; they identify problems and needs, then use assets and apply innovations to address these needs. Different stakeholder organizations and institutions bring various skills, contributions, and initiatives to problem solving.

The CRFS project team recognized early the value of collective impact as a perspective through which to observe the impact of food system interventions and as a strategy for some of the project's own direct actions. The team adopted Kania and Kramer's (2011, 36) "five conditions of collective success" to understand particular influences on food system change within the CRFS project cities:

- A common agenda for change shared by all participants
- Shared measurement systems agreed upon by all
- Mutually reinforcing activities performed by specific participants under a guiding overall plan
- Continual communication among all participants to share information, report on progress or challenges, and develop mutual trust
- A separate organization acting as the backbone for the entire initiative

The US community food security movement has reflected some aspects of collective impact without adopting the term. Activists realized and acted on the knowledge that the broad societal goal of achieving consistent and universal food security is multidimensional, with linkages across several levels. With many organizations and individuals engaged in the work, efforts to communicate were important. For almost two decades, the Community Food Security Coalition acted informally as the movement's backbone organization, particularly during periods of advocacy before the periodic adoption of new farm bills by Congress. After that national organization ceased operations in 2013, certain collective impact approaches were locally adopted by place-based food policy councils composed of members representing different food system components working under a common agenda; the Los Angeles Food Policy Council and the Milwaukee Food Council are two examples highlighted in this book (see chapters 10 and 12). Also, one direct outcome of the CRFS project, the Cooperative Institute for Urban Agriculture and Nutrition in Milwaukee, was created and announced as a collective impact initiative.

In this book we also mention where and how community-driven efforts to work toward food system change can break down. Collective impact efforts may be impaired by competition or distrust between organizations, lack of coordination or leadership, limited resources, resistance from businesses and others that profit from the status quo, restrictive policies and intransigent political leaders, and indifference from consumers. In some circumstances advocates and organizations may have to be content with incremental changes that might lead to a greater effect in the future—such as helping to improve a single neighborhood, giving a few kids a job and job skills, or showing a group of people how to garden.

Innovation

A brief explanation is needed for what *innovation* denotes in this book—and for the CRFS project as a whole. Although the project did not adopt a specific definition of innovation as a guidepost, its understanding of when and how a food system action or outcome represents a social innovation largely reflects the belief of Phills, Deiglmeier, and Miller (2008): the essential elements of an innovative act involve a creative process and its product, but also the diffusion or adoption of the innovation into broad use and the ultimate level of social value caused by the innovation.

Missing from this conception is the idea that a prerequisite for innovation is a prior failure, because the potential for innovation often arises from lessons learned through failure (Patton 2011). In a broad sense the concept of community food systems represents an innovative reaction to the apparent failure of the decades-old paradigm that the corporate food system is capable of feeding everyone adequately. With the failings of that paradigm (as outlined in this chapter), the community food system movement has gradually assumed the identity of an alternative paradigm, with its own set of generally accepted beliefs and guiding principles. Thus, innovation within the movement largely takes the form of new and different project-scale actions implemented within the alternative paradigm. These actions were the innovative practices sought out by the CRFS project.

Individual actors in community food security, aware that they are part of a movement, look to the past work of colleagues for help and guidance. This often involves the sharing of best practices, the dissemination and adoption of innovations that have worked elsewhere. This approach assumes that the characteristics of the situation to be addressed are similar to and consistent with those from which the best practice originated (as an innovation). Innovation occurs through other channels as well. If a best practice is deemed not applicable or appropriate in a certain context, an original approach must be generated; or a particular practice might simply be considered outdated in some way. In perhaps the most prevalent generator of innovation, a situation to be addressed simply has no precedent and thus no proven approach to adopt.

As connoted by the official title and informal tagline of the CRFS project—"identifying innovations and promoting successes"—the project team sought to highlight innovative food system practices and credit the individuals and organizations who design new and creative approaches to community-targeted food system actions. In some examples portrayed in this book, the innovations were created through the direct use of CRFS project resources in community settings within one of the seven project cities.

A FRAMEWORK FOR FOOD SYSTEMS

Our original proposal to the USDA that initiated the CRFS project included an impossible objective: using a systems analysis approach to create an overall framework and a conceptual model of community and regional food systems. These systems are highly complex, dynamic, interlinked with broader influences and processes, and variable in time and space. They are not easily deconstructed into their components, as in the process of systems analysis. Nonetheless, we have built several iconographic frameworks during the course of the project. These have been quite useful for labeling the components of and the influences on food systems, organizing our community interactions, and creating a common language for discussing concepts and exchanging ideas among researchers, advocates, and practitioners.

We used our final framework as an organizing tool throughout this book. The evolution of its graphic representation is emblematic of the evolution of our thinking and so is briefly reviewed.

Our original USDA proposal contained a text-based description of the elements of a food system. Each of these elements included a description of what aspects or dimensions are important for a community or regional food system:

- Land suitability
- Land tenure and land economics
- Production systems, production, and processing infrastructure
- Transportation and logistics
- Marketing, markets, and food distribution
- Business models and management
- Capital and labor
- Community and cultural relations
- A suitable legal and political environment

These elements were rendered into a graphic representation early in the project. At one year into the project, the food system components in the diagram had expanded to thirteen; members of community organizations we worked with insisted that a representation of a food system they cared about had to explicitly include education, nutrition, and health perspectives. Lenses were used to show how complex food systems could be viewed in multiple ways.

Through several more iterations of graphics, we ended up with our double-wheel diagram (fig. 2). The inner circle of the right wheel is a food system supply chain

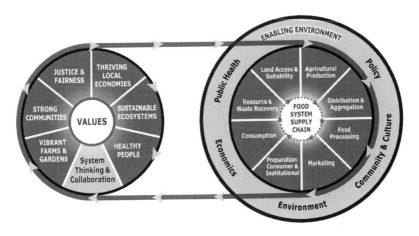

Figure 2. Final diagram of the CRFS framework. (Courtesy of CRFS)

or value chain. Similar diagrams have been used in multiple publications, and the concepts have been discussed in the context of local and regional food systems by Bloom and Hinrichs (2010), Lev and Stevenson (2011), and Hardesty et al. (2014). These food supply chain elements operate in a broader "enabling environment," a set of proximate influences on community food systems within a societal context. It is shown as distinct from the inner food supply chain so that we do not imply a one-to-one relationship between specific outer- and inner-ring elements. For example, policies at multiple levels affect every component of the food supply chain.

The figure explicitly recognizes a set of values that drive community food systems. This concept is depicted as though the values were linked to a chain or a belt driving the food supply chain. This depiction has been quite useful for describing the distinction between corporate and community food systems. For example, we could say that a corporate food system has only one cog—profit (roughly equivalent to "thriving local economies")—driving the food supply chain, whereas a community system has most or all of the cogs shown in the graphic.

The dark shaded values shown in the left wheel are derived from Abi-Nader et al. (2009). This seminal work from the Community Food Security Whole Measures Working Group provided the framework for some of our CRFS training and is one way of utilizing the resulting Internet resources now maintained by the Local Food System team of the University of Wisconsin Extension (n.d.). We added a seventh value, systems thinking and collaboration, to the original six values in recognition of the need for a broad perspective on community food systems to foster innovation and promote cooperation, in keeping with the intent of collective impact concepts.

Our community food systems diagram became the organizing structure for this book. Although the book's final iteration does not exactly parallel the eight elements of the food supply chain and the seven value components, we have attempted to cover all of them. Chapters 2 through 8 correspond to elements of the food system supply chain, and the remainder of the book examines aspects of community and regional food systems values.

The Five As of Food Security

STEVE VENTURA

- **Available.** The FAO's definition of availability refers to the supply side of food production and distribution. In the context of community food systems, the term *availability* is used to connote food produced, processed, and distributed within a region. Thus, a community food system definition includes the nonfood benefits of local production and processing, including retaining revenues in the local economy, informing consumers where their food comes from, and generating other community benefits. The CRFS project did not explicitly define *region*, based on the observation that an appropriate "foodshed" (Kloppenburg, Hendrickson, and Stevenson 1996) may vary by commodity. For example, in Wisconsin the region for highly perishable and difficult-to-ship products such as sprouts and fresh greens may be a single Milwaukee neighborhood, whereas the region for carrots or potatoes may be many hundreds of square miles, incorporating the state's Central Sands vegetable production region, where it is easier and cheaper to produce these commodities.
- **Affordable.** For decades US federal food policy has been oriented to the provision of cheap food for the masses. Although this model has made the cost of food a relatively small part of most people's income, food policies have favored and subsidized a small number of crops and animals processed into food that is high in calories and low in nutritional value. As a result, healthy food such as fresh fruits and vegetables are expensive,

and organic and local produce are often more so. Even middle-income consumers may eschew these healthy options because the costs are higher. Public food and nutrition programs such as the Supplemental Nutrition Assistance Program (SNAP, better known as food stamps) alleviate the affordability gap for the lowest end of the income spectrum. Local incentives such as vouchers and additional subsidies can be used to direct purchases toward more nutritious food.

- **Accessible.** The term *food desert* has become shorthand for areas lacking access to supermarkets and other sources of fresh nutritious food. However, this term is considered disparaging by food advocates who work in such areas, both because the areas often have a glut of food sources such as fast-food restaurants and gas station markets and because it oversimplifies a complex problem involving food outlets, transportation, crime, city development and investment patterns, and other factors. Low-income people may not have access to their own automobiles, so they need access to healthy, nutritious food in stores and restaurants in their neighborhoods or effective public transportation to full-service supermarkets, although the latter option restricts the volume of food that can be purchased at one time. Several innovations could alleviate the lack of access to supermarkets, including farmers' markets, community supported agriculture (CSA) shares and market baskets from nearby farms, delivery trucks, and corner-store initiatives (described further in chapter 6) that help small markets offer a wider selection of fruits and vegetables. Many farmers' markets now have the capacity to accept electronic benefit transfer (EBT) cards, the medium used for SNAP purchases.
- **Appropriate.** Food should be appropriate from multiple perspectives. From a health standpoint, it should be nutritious and safe—free of pathogens and contaminants such as deleterious levels of pesticides and heavy metals. Food should be culturally relevant. In addition to religious strictures on allowable or banned foods, the cuisine of a community as it has evolved from customs and traditions strongly influences what foods are consumed. New foods may be wholesome and nutritious, but people may not try them if they are taboo or too strange. Food should also be appealing. Fresh vegetables and whole grains in particular have an often-deserved reputation for being bland, especially compared to sweet, salty, and greasy common fare. Preparation skills are a necessary component in mak-

ing food appropriate, along with an appreciation for preserving nutritional value during preparation.

- **Acceptable.** *Sustainability* and *triple bottom line* are overused terms, but the core concept remains useful. Our food system should be environmentally benign, useful for consumers and communities, and profitable for everyone involved from farm to fork. Without environmental safeguards, the long-term security of our food supply is threatened in multiple ways, including soil loss from erosion, depletion, and contamination; the loss of genetic diversity in our crop plants; and the loss of pollinators. In the United States the most important aspect of food security is remaking the food system so that it benefits communities, not just large corporations and megafarms. Creating a community orientation that is compatible with our free enterprise system and market economy is an ongoing challenge and a critical part of food security.

CHAPTER TWO

Land Tenure for Urban Farming
Toward a Scalable Model

NATE ELA AND GREG ROSENBERG

In this chapter attorneys Nate Ela and Greg Rosenberg describe the challenges of secure access to land for urban growers, discuss community land trusts as a possible tool for maintaining land for food production in cities—the "land access and suitability" component of the food system supply chain—and introduce several practical models for securing land for urban farming.

In cities around the United States people are looking to urban farms for a wide range of benefits, from providing fresh food and supporting healthy eating habits to teaching job skills and offering access to nature. Urban farms are increasingly seen as a possible engine for economic development, whether as a source of income for full-time farmers, of raw materials for value-added products, or of additional income for growers who also hold evening jobs or off-season gigs.

Yet none of this can happen without land on which to grow crops. In a twist on the popular slogan "No farms, no food," urban growers and their advocates have come to realize "No land, no farms." A wide range of people—from individual growers to the mentors, foundation officials, university researchers, urban planners, and policy makers who would like to see them succeed—have been grappling with questions related to land tenure for urban farming. What models for land ownership and access can best support planning for and investing in urban farms? What models allocate scarce resources not only efficiently but also in ways that promote equity and engagement with the communities that will be the urban farmers' neighbors and customers?

This chapter explores these questions, focusing primarily on land tenure models that can offer urban growers free or low-cost land and that hold the most promise for the sustainable growth of urban farming. The success of such models will depend on cooperative efforts with public and private landowning entities that can make land available at no or minimal cost, and with public and private funders that can cover some of the costs of land acquisition, soil remediation, and basic infrastructure. Success will also turn on the ability to build community engagement into the rules by which urban land is made available to growers. Historical patterns of investment and disinvestment have created areas in US cities where vacant land is abundant and relatively cheap. But it would be a mistake to simply assume that land can be leveraged as a resource for farming without ensuring that it makes sense to the people who live in these neighborhoods—often communities that have a long experience of structural racism.

Although urban agriculture takes a wide range of forms, we focus here on land tenure models that can best support ground-based, outdoor growing of commercial crops. Such growing practices, compared to growing on rooftops or indoors, are more likely to yield the full range of community benefits mentioned above. Also, because community nonprofit organizations or individual growers may have less access to capital than rooftop or indoor growers, the cost and availability of land is an even more pressing constraint.

THE CHALLENGE OF SECURING AFFORDABLE LAND FOR URBAN FARMING

The high cost of urban land compared to rural land poses a major problem for would-be urban farmers. Unlike rural farmers, they are competing for land with many other potential uses, which creates inflationary pressures on land prices. In Wisconsin, for example, cropland rents for rural land averaged $228 per acre in 2015 (USDA National Agricultural Statistics Service 2015). This is a small fraction of the price an urban farmer would pay for an acre of tillable land at market rates in Milwaukee or other cities. Yet food grown in cities must remain price-competitive with food grown in rural areas. Few if any crops can be sold at prices that would cover the higher land costs, and urban growers cannot simply add a premium to reflect the value of the contributions they make to their neighborhoods. This is the well-known issue of *positive externalities*: beneficial side effects of economic activity that might be underproduced when a good or service is valued at market rates.

This squeeze between the price of produce and the cost of land drives many urban farmers to look for free or low-cost land, which often leads them to the parts of cities where the market value of land is most depressed. These neighborhoods have the all-too-familiar histories of disinvestment, including white flight to the suburbs. They are neighborhoods from which industries too have fled, moving jobs overseas and often leaving contamination in their wake. Such areas are often beset by increased crime, abandoned and neglected buildings and infrastructure, and cuts to public services. These factors moderate the pressures that would drive up the price of scarce urban land in healthier economic, community, and environmental circumstances. These areas present both challenges and opportunities different from those faced by farmers in rural settings.

The Need for Tenure Protection

The prospect of secure, long-term land tenure creates opportunities previously unavailable to urban farmers. For example, they can consider applying for organic certification, a process that often takes at least three years to complete. Also, long-term land tenure justifies new levels of investment in soil remediation and infrastructure such as hoop houses and irrigation systems, which are often not feasible for growers operating on a year-to-year basis.

High land costs and market pressures stand out as the greatest obstacles to long-term land tenure for urban farmers. Other forms of development can yield far higher rates of return for investors, which forces urban farmers to justify why they deserve access to otherwise undeveloped land at below-market rates.

A long-term tenure may not be the ultimate goal for all farmers or every neighborhood. Long-term leases or outright ownership can be the best fit for nonprofit agricultural organizations that provide farmer training, for neighborhood organizations that include community gardens or farmer leases as part of their mission, or for commercial ventures run by experienced farmers. But new farmers graduating from training programs may need a few years to experiment with business models or recognize the reality beyond the romance of urban growing. Because many may not continue to farm beyond the first few years, short-term leases could be a good match.

KEY CONCEPTS

Access to land for urban farming entails several considerations, including land markets, property taxes, the mechanisms for land tenure protection, and collaboration with others around land tenure.

Shielding Land from Speculation and Desperation

If affordable land is essential for commercially viable, community-engaged urban farming, the question becomes how to protect affordability long-term. This means ensuring that urban farms are not displaced by rapidly rising prices in a speculative real estate market and that land in communities with stagnant or declining values is not allocated willy-nilly in desperation. In both cases, the struggle is how to ensure that space is available to projects that are rooted in and beneficial to the surrounding communities.

In the last several decades, housing and environmental advocates have developed land trust models to ensure that community priorities—places to live and places to enjoy nature—are not displaced by speculative market forces. Open space and conservation land trusts have focused on protecting ecologically valuable land at the urban fringe, and community land trusts have sought to protect housing affordability in cities and suburbs. Both models are increasingly being brought to bear on the question of how best to protect land for urban agriculture.

Adapting these models to conserve productive urban farmland often involves ground leases from a nonprofit land trust to a grower. The community land trust owns or holds rights to the use of a parcel and essentially rents the land to a farmer. The rate will depend on what is necessary to maintain the property, including what can be subsidized through public or private philanthropy, property tax breaks, and the overhead of the trust. This arrangement presents the concern that growers may lose out on certain opportunities for profit because they do not own the land. They will not have the opportunity to gain from selling their land at market price because they did not purchase it at the market rate in the first place. For many urban growers the outright purchase of market-rate land is likely to be out of reach and effectively not an option.

Considering the Ownership Trap

In many cases outright ownership may not make the most sense for urban farmers. Even if an urban farmer sought to purchase an undeveloped parcel with a price so low that financing was not necessary, there would still be the upfront costs of securing the title and paying transaction fees. Outright ownership could also create property tax obligations, which might be mitigated by leasing from a nonprofit land trust, as discussed below. Then there are the costs of municipal services, insurance liability concerns, and the difficulty of selling the property if necessary.

Added together, these costs and liabilities could make ownership a trap. Rather than assuming that ownership is either the gold standard or something to be avoided, the key is to determine the degree of security of tenure that best matches a particular grower's goals and experience.

Paying Attention to Property Taxation

From the perspective of the grower and landowner, it is always best to reduce or eliminate property taxes to protect the affordability of urban farmland. Of course, the taxing authority may have a different perspective, but there are ways to structure land ownership to reduce property tax burdens without asking for special treatment from the local municipality. For example, where agricultural land may be assessed at its use value, an agricultural easement could reduce the property tax burden. Or a nonprofit land trust may be able to hold tax-exempt land that could be leased to beginning farmers who are part of an educational incubator program.

Property tax assessors may be able to provide favorable assessments for land used for open space or agriculture. Yet individual growers often don't know whom to approach in these agencies or don't understand the processes of securing land and seeking favorable tax treatment. For example, Wisconsin has an agricultural use value property tax provision, but to date local assessors have not extended this reduced valuation to urban agricultural land.

Such property tax considerations, then, may be a significant determinant of who should own the land. Each form of ownership carries different implications for property taxation, along with other challenges and benefits (table 1).

Securing the Appropriate Type of Land Tenure

Land tenure must allow the growers to recoup their investments. Although pop-up business models are increasingly popular in the restaurant and retail sectors, it is a difficult model for urban growing. Access to land for a single season requires growers to only make investments they can pick up and move easily. For growers to invest more deeply in a piece of land—by preparing it for growing, installing infrastructure, or seeking organic certification—they require contractually assured access over several growing seasons. One might expect that the greater the level of initial investment, the longer a grower's time horizon will be.

Appropriate land tenure also entails matching organizational goals to specific parcels of land and matching farmers' goals to the land they lease or own (table 2). Most nonprofit agricultural organizations will seek long-term

TABLE 1. Benefits and Challenges by Type of Landowning Entity

Landowning Entity	Benefits	Challenges
Public	Exempt from property taxes.	Difficulty in securing long-term ground leases; change in the political winds could cause loss of land for farming.
Nonprofit	May be exempt from property taxes for noncommercial and incubator farms; commercial farmland could be assessed based on ground rents.	Negotiations with the local assessor can be thorny, particularly in inflated markets or in cities struggling to generate property tax revenues.
Private	The least red tape, particularly when the grower owns the land directly.	No benefit from a property tax perspective except in jurisdictions that have a use value property tax system that includes urban agriculture.

tenure, either in the form of long-term ground leases or outright ownership. Apprentice farmers who have not yet established an ability to successfully farm over time will be better candidates for short-term leases. These could include performance-based opportunities for renewal, which would effectively allow the growers to establish their own land security. Experienced urban farmers are more likely to seek land tenure that allows them to build a long-term business and potentially pass down the land to successors, whether they are family members or business partners.

Interacting, Specializing, and Working in Teams

Urban growers seeking access to affordable land often turn for help to government officials at the city and county levels. City officials and land bank managers can help identify vacant land that could be available for farming. In many US cities and counties, land banks have been established as quasi-governmental authorities to

TABLE 2. Desired Tenure by Tenant Type

Tenant Type	Key Concerns	Tenure Type
Apprentice growers	Because they are new to farming, it is uncertain whether they will be successful long-term.	A short-term ground lease with a renewal clause based on performance enables farmers to earn tenure while protecting the landowner if the farmer is unsuccessful.
Experienced growers	Because they have a track record of success, the risk of farm failure is significantly lower than for apprentice growers.	A successful track record provides more confidence to landowners to provide long-term leases. Depending on market conditions and financial resources, a grower may opt for outright purchase.
Nonprofit agricultural organizations	As entities that plan on being around for a long time, nonprofits will desire the longest possible security of tenure for their land.	Outright ownership or ninety-nine-year ground leases may be the best fit, unless short-term tenure better aligns with the goals for specific parcels of land.

manage their inventory of surplus land. From the perspective of city and county officials, it is easier to work repeatedly with a single organization such as a land bank or land trust that represents multiple growers rather than beginning the process anew with each grower.

For small organizations, it can be difficult to develop and sustain the expertise required for complex real estate matters. Staff members are often generalists, responsible for multiple subject areas, who cannot take the time necessary to learn the intricacies of real estate law. In this context it makes sense for urban farmers to seek the services of organizations that can afford to specialize and develop relationships to negotiate effectively with city and county officials. Examples are described later in this chapter.

In short, urban farming is a team effort. Even when growers appear to be raising crops all by themselves, in most cases they are dependent on the support of others. Securing affordable land to farm in the first place requires working with real estate professionals, accountants, lawyers, nonprofit organizations, and government officials. To develop a food system in which urban farmers can focus on what they do best, we should not ask them to play every position on the team or be masters of all domains. Enabling partnerships between urban growers and trusted specialists who can work with government officials should maximize the benefits that urban farms can bring to a city's landscape.

Program Design: Eight Strategic Questions on Landholding for Urban Farms

NATE ELA AND GREG ROSENBERG

1. HOW WILL LAND BE SECURED FOR FARMERS?

Although this is the central question to be answered by a land tenure model, we do not expect that there will be a single answer. As suggested in the previous section, land may be secured differently for farmers with different levels of experience.

Before land is secured for particular farmers, however, there is the question of how to protect land for agricultural use. This could mean transferring publicly or privately owned land into a land trust, which then provides leases to individual farmers or urban farming organizations.

Whether land is leased by a land trust or a public agency, it makes sense to have different terms for different types of farmers. Nonprofit urban farms could be eligible for long-term leases—up to ninety-eight-year renewable leases for the most well-established organizations. Such leases would ensure long-term agricultural use and provide security to urban farms that are committed to being an ongoing resource for a neighborhood.

For individual farmers, a renewable short-term lease could have performance measures negotiated by the farmer and the leasing entity, with input from community members. Farmers could thus work their way into long-term security of tenure by demonstrating their ability to pay the (below-market) rent and provide community benefits.

2. HOW WILL LAND BE MADE AFFORDABLE?

If urban farmers are to have any hope of sustained success, their costs for land access should be roughly on par with those of rural farmers. Thus, one reasonable target for affordability would be for urban farmers to devote the same percentage of the cost of input to land as rural farmers do. For rural farmers, this proportion will depend on the crop, whereas urban farmers will be more likely to have a more intensive and diversified growing strategy.

3. HOW WILL LAND BE USED?

What type of land is appropriate depends on how growers plan to use it, of course. Will they grow in greenhouses, hoop houses, or outdoors? Will they be growing flowers, herbs, or vegetables? Will they set up composting facilities? Land use will depend not only on the growers' desires but also on zoning and other regulations.

4. WHO WILL BE THE FARMERS?

A land tenure model must be responsive to different types of farmers. These types include job trainees working on nonprofit urban farms, new growers testing their business models on incubator farms, and independent growers with just a few years or decades of experience. A tenure model can also help encourage community-engaged urban agriculture by minority-run firms and by prioritizing access to land for farmers who will grow in their own neighborhoods.

5. WHAT TYPE OF SUPPORT WILL FARMERS NEED TO BE SUCCESSFUL?

Support will vary widely based on the experience of the farmers, issues relating to the land, and challenges in accessing the local market for their produce. For land-related issues, farmers may need support for soil remediation, the installation of infrastructure (water and electricity), the construction of agricultural buildings, the negotiation of favorable property tax assessments (if they are the landowners), and zoning changes (in some cases). The support of a team of people and organizations is usually required to address all these issues.

6. HOW IS SUCCESS DEFINED? WHAT EXPECTATIONS ARE REALISTIC?

In defining a system for land tenure, people must grapple with what a successful urban farming sector looks like. Although nonprofit urban farms have been demonstrably effective sites for youth programming and job training, most cities have not seen a large number of small, for-profit urban farms that create many well-paying jobs. If communities or government officials expect urban farms to be a major vehicle for short-term job creation, those expectations may be unrealistic.

A successful land tenure model should support land remaining in agricultural use for a period in which urban farmers can test out for-profit and nonprofit business models. It will take time for farmers to learn which business models provide an acceptable mix of economic return and community benefits. Along the way some farms will fail. This is normal with small businesses; the Small Business Administration (2012) has found that only about half of small businesses survive the first five years. Such failures are not a sign that land should not be preserved for agricultural use; rather, a successful land tenure model would quickly provide access to a new grower.

7. WHO SHOULD BE THE LANDHOLDING ENTITY?

Different types of entities could hold land for urban farms, including government agencies, land banks, agricultural cooperatives, or even private firms. Cities have large parcels associated with churches, corporate headquarters, educational institutions, and public agencies that may include unused acreage appropriate for urban agriculture. Cities such as Oakland (California), Portland (Oregon), Madison (Wisconsin), and Philadelphia have conducted inventories to determine where such opportunities exist.

Around the United States people are increasingly looking to land trusts as an entity suited to holding land for urban farms and gardens. However, property tax issues are an important consideration in whether a land trust itself should hold the title to the land or instead manage land held by public entities; market-rate property tax assessment can make land unaffordable, even for a nonprofit land trust. (For this reason, the Athens Land Trust in Athens, Georgia, opted not to own urban farmland; unfavorable property tax treatment would have yielded a full market-rate assessed value for the land, regardless of long-term restrictions placed on it.)

8. HOW WILL THE LANDHOLDING ENTITY RELATE TO COMMUNITY MEMBERS?

Whether land is held by a nonprofit land trust, a government agency, or some other entity, the relationship between the landholder and community members will inevitably be a key question. Are community members included as the board members of a land trust, and if so, how? Are they consulted by the decision makers in a city landholding agency or a county land bank, and if so, by what process?

THE ROLE OF NONPROFIT ORGANIZATIONS IN URBAN FARMING

As in other areas of community economic development, the nonprofit sector has a special role in kick-starting urban farming. Urban agriculture is a relatively low-cost approach to community revitalization, and the cost of investment is lower than in other forms of redevelopment. At least in principle an urban farm can be built more quickly and cheaply than housing or mixed-use development. In practice, of course, the fact that urban farming business models are still being tested means that they can encounter delays in raising capital and satisfying regulatory requirements.

Nonprofit urban farms can therefore be seen as effectively paving the way for subsequent private-sector development. Beyond farming, this is the historical role that nonprofit groups have assumed in community economic development. For-profit housing developers, for example, may be hesitant to enter a neighborhood in which there is a high perception of risk. They will wait until nonprofit groups have worked out the regulatory wrinkles and have proved that the demand is sufficient to justify investment.

Affordability

The parallel between housing and food production suggests why nonprofit organizations—in the form of land trusts, training sites, and incubator farms—could be an essential provider of affordable urban farmland. In "unaffordable" housing markets, by analogy, the provision of affordable housing is not possible without some subsidies, nor can affordable housing be maintained in the face of a rising market without some kind of controls, such as resale restrictions, that ensure long-term affordability.

Unlike housing, urban farmland has no standardized definition of affordability. Housing affordability is most frequently described as a percentage of gross income; that is, housing is considered affordable if no more than one-third of gross household income goes toward housing-related expenses. In the United States targeted household income is expressed as a percentage of area median income, which ranges from 30 percent for low-income households to 120 percent for moderate-income households in hyperinflated markets such as the San Francisco Bay Area. For urban farmland there is no comparable framework.

Instead, affordable urban farmland is often described simply as land that is "free or cheap," with little description beyond that. Even "free" land is rarely free, as it almost always needs some level of soil remediation or installation of infrastructure—and then there are always transaction fees. Affordability is usually defined on a case-by-case basis relative to the specific parcel of land, the need for remediation and infrastructure, the crops being grown, and the net revenues that a grower would need to generate.

In urban land markets where space for farming is unaffordable because of the current revenue models for urban farms, some measure of subsidies could be justified to make land available for growers to learn basic skills and test their emerging business models. As the market develops, the most skilled farmers may be able to afford land at rates similar to those in nearby rural production areas (outside hyperinflated real estate markets), but some controls would probably remain justified to provide space for new entrants to urban farming and to ensure that all neighborhoods enjoy the community benefits that urban agriculture makes possible, even if land values eventually rise to a level that would otherwise preclude farming as an economically viable land use.

Nonprofit Groups as Partners, Not Predators

Because low-income minority neighborhoods are so often characterized by what they lack, it is easy to ignore what they have. For nonprofit organizations these neighborhoods are fertile ground for planting new initiatives and for supporting or expanding programming. Nonprofit leaders often act with good intentions, aiming to support community revitalization, but there are also pressures to respond to funders' expectations that programs will target the "most needy" communities, which can be used as testing sites for developing scalable or replicable interventions. At the grandest scale, whole cities—such as Detroit, or New Orleans after Hurricane Katrina—have been imagined as laboratories for experimenting with the types of interventions favored by donors and the nonprofit organizations they fund. Community residents are sometimes left feeling more like lab rats than partners.

Other chapters in this book discuss how urban agriculture can potentially be the sector in which people resist—or reproduce—the types of oppression that have been woven into the history of the United States. In the following sections, we seek to identify how land tenure models can be structured to ensure that urban farming nonprofit groups are partners of the communities in which they work rather than predators.

THE ROLE OF LAND TRUSTS IN PROVIDING AND PROTECTING AFFORDABLE LAND FOR URBAN FARMING

In the last thirty years the land trust has emerged in the United States as a preferred model for holding land for community gardens and urban farms. This reflects the convergence of two trends: the creation of specialized open space land trusts to conserve land for community gardens, and the moves that some community land trusts have made to promote urban agriculture.

There is a distinction between open space land trusts and community land trusts. Open space land trusts focus on the protection of land and generally do not have structures in place to manage lands that are being used productively or to foster community-based governance. Community land trusts, in contrast, acquire and hold land for the benefit of a community and generally have a tripartite board structure that includes seats dedicated to beneficiaries of the trust (usually people who live in housing held by the trust), residents from neighboring communities, and people with the necessary expertise or organizational connections.

There are, however, movements toward open space land trusts. Openlands, the regional open space land trust in the Chicago area, is focused on the ecological potential of farmland that is used as a buffer for conservation lands. Also, the Trust for Public Land (n.d.) has recently developed a "working lands" initiative. Meanwhile, a move is under way in the open space land trust community to promote "community conservation" initiatives (Aldrich and Levy 2015).

To illustrate the different forms that such trusts can take, we provide two examples. In a later section we examine the model developed by Chicago's NeighborSpace land trust.

Open Space Land Trusts: New York Community Garden Land Trusts

In 1999 the administration of New York City Mayor Rudolph Giuliani announced a plan to auction off more than one hundred pieces of city-owned land that were home to community gardens. Gardeners and their allies mobilized in resistance to the plan, organizing demonstrations and filing lawsuits (Brooklyn Queens Land Trust n.d.). In 2002, after a negotiated settlement of the lawsuits with Mayor Michael Bloomberg's administration, sixty-nine gardens were purchased and held in trust by the Trust for Public Land. The New York Restoration Project, a nonprofit organization founded and funded by the entertainer Bette Midler, took ownership of dozens more gardens.

In the years since, the Trust for Public Land established three local land trusts to hold and manage the gardens: the Manhattan Land Trust, the Brooklyn Queens Land Trust, and the Bronx Land Trust. The board of each land trust is a mix of community garden leaders and staff from New York City nonprofit organizations. The New York Restoration Project (n.d.) has taken on a broader mission to provide green space to underserved areas of the city and is led not by gardeners but by a range of New York philanthropists, businesspeople, and civic leaders. Some of its sites have been renovated with support from corporations such as Target, and they have reduced the space available for community-managed gardens in favor of tidy pocket parks.

Community Land Trusts: Southside CLT

Since the 1960s the community land trust (CLT) has provided support for affordable housing in the United States. In the housing sector the CLT model combines two basic innovations in land governance. The first innovation is to separate the ownership of a home from the ownership of the land on which it is sited. A homeowner in a CLT holds the title to the home, but the title to the land is held by the nonprofit CLT, which then provides a ground lease to the homeowner. This arrangement allows CLT homeowners to build equity in their homes while ensuring affordability to subsequent homeowners by limiting the amount by which the land's sale price can increase.

The second innovation is the tripartite structure of the CLT board, whose members are generally split evenly among homeowners, members of the neighboring community who are not part of the CLT, and the types of community leaders (e.g., representatives from other nonprofit groups or credit unions) typically associated with a nonprofit board. This board structure ensures that the CLT remains committed to balancing the interests of its members and the community.

In recent years CLTs have taken on three main roles in support of urban agriculture. First, some CLTs that were formed to support affordable housing have begun to hold land for community gardens and urban farms. An example is Troy Gardens, a project of the Madison Area Community Land Trust, which incorporated community gardens and an urban farm alongside affordable housing (Rosenberg 2007). Second, some of these housing-focused CLTs have provided programmatic support for urban agriculture other than taking on ownership of land. Third, a few organizations have been founded as CLTs that are exclusively focused on urban agriculture, adapting some of the techniques for community-focused governance from the CLT model developed in housing. Here we focus on the Southside Community Land Trust (n.d.), which focuses on preserving land for community gardens and urban farms.

Southside CLT holds the title to sixteen community gardens in Providence, Rhode Island, and provides programmatic support (such as arranging for bulk purchases of organic fertilizer) for these gardens as well as twenty-five gardens owned by other organizations in its network. Southside CLT differs from other CLTs by holding land for gardens and farms rather than for affordable housing. However, like traditional CLTs it has built community representation and engagement into its governance structure. The gardeners themselves must elect 51 percent of the board members (Yuen 2012, 36–37).

Southside CLT has drawn on land conservation tools developed in suburban contexts to raise revenue and provide additional levels of protection for its community gardens. It sold the development rights to a number of its gardens to the Rhode Island Department of Environmental Management and used the revenue to offset some of the cost of acquiring the land. Selling development rights to the state also helps to increase security of ownership, because a developer would need to acquire both the title to the garden and the development rights to get the property. The use of a state open space bond to finance the purchase of development rights further restricts potential land uses and adds to security of tenure (Yuen 2012, 22–23).

In addition to protecting land for community gardens, Southside CLT manages two commercial farms. City Farm is a three-quarter-acre commercial urban farm in south Providence that began in 1986. Urban Edge Farm is a fifty-acre farm in nearby Cranston, Rhode Island; its mission is to support seven new farmers who collaboratively manage the land. The land for Urban Edge was purchased by the state in 2002, pursuant to the state's Open Space Preservation Act. The site, which was formerly a dairy farm, is now owned and protected by the Rhode Island Department of Environmental Management (n.d.), which leases it to Southside CLT for one dollar a year (Ewert 2012, 97). About twenty of the fifty acres are cultivable.

Southside CLT initially operated its own CSA farm at Urban Edge, but within a few years it became clear that production would not cover the significant staffing costs (Ewert 2012, 91). Through Urban Edge Farm, Southside CLT now teaches farming practices to new farmers, rents them farm equipment, provides compost and fertilizer, and plows the land once a year (Snowden 2006). After going through training, the beginning farmers can rent up to two acres of land at below-market rates. These farming businesses are owned and operated by people who have experience in farming but weren't able to buy or rent land on their own at market rates. They sell through CSA shares, directly to institutions, and through growers' cooperatives. The terms of Southside CLT's lease with the state are meant to prevent competition with nearby farms, and thus bar on-farm sales or nonfarm businesses.

THE CENTRAL SERVER MODEL: A SCALABLE APPROACH TO URBAN FARMING

The central server model is an approach developed within the community land trust movement to facilitate the rapid citywide scaling of permanently affordable housing while striking a balance between local control and economies of scale. The model was first introduced in 2009 in Atlanta (*PD&R Edge* 2012; Schneggenburger 2011) and soon thereafter in New Orleans (Khanmalek 2013). Its supporters hoped to spur growth in the number of neighborhood-based community land trusts by creating a central entity that would provide a variety of technical services: accounting, development, and real estate transactions; negotiating with funders and lenders; and other services that require more expertise than a neighborhood-based organization can easily muster.

From the experiences in Atlanta and New Orleans, affordable-housing advocates have learned that the burdens placed on a central server can be onerous from legal, political, financial, and community relations perspectives. We do not have space in this chapter to evaluate the effectiveness of the central server model for affordable housing, but it is worth noting that affordable-housing community land trusts continue to adopt and adapt the model (Dudley Street Neighbors 2015).

In the context of urban farming, in which transactions are more straightforward because they do not involve housing or residents, we believe that the central server model holds significant promise. Successfully implementing this model in this context depends on striking a balance between local control and economies of scale. An appropriate architecture would involve a web of neighborhood-based satellite organizations served by a citywide central server organization. The central server provides a suite of services to the satellites and to the farmers to whom the satellites provide land.

The Role of the Central Server in an Urban Farming Context

A central server can do the "heavy lifting" that is beyond the ability of small, neighborhood-based organizations, particularly while they are focused on starting up urban farming projects. With expertise in land use and real estate transactions, a central server can negotiate with the local government to secure publicly owned land for agriculture, obtain favorable tax treatment, and gain access to city services to provide the necessary infrastructure to gardens and farms. In addition, a central server could help provide training and technical support to its satellite organizations. Providing such services in group settings is less costly and creates opportunities to build connections between satellite organizations.

A central server can also provide a single point of connection to funders. This can increase the collective leverage of neighborhood organizations beyond what they could accomplish individually, and it would reduce overhead for funders by packaging what might otherwise be numerous similar grant applications. (Of course, satellite organizations may also seek funding for their own operations.) A central server may also have access to officials and decision making in city government that is beyond the scope of a small organization.

The term *server* is the key to understanding the model. A central server exists to serve satellite entities and the neighborhoods they serve. The staff for a central server must be skilled at working well with others and protect the central server's role as a universally trusted entity by studiously avoiding turf battles or picking favorites. This is not simple, especially in cities where local elected officials have a strong say in how projects are developed in their districts.

The Role of the Neighborhood-Based Satellite Entity

Satellite organizations serve as the voice of the community. They may be existing nonprofit organizations (e.g., community development corporations or community land trusts), new start-ups, or more informal entities. No matter what form they take, they must be able to speak credibly on behalf of their neighborhoods and ensure that land use decisions are in the best interest of the residents.

A central server frees neighborhood-based satellite organizations from having to deal with real estate transactions, infrastructure installation, and the negotiation of favorable property tax treatment. The satellites can therefore focus on the critical work of governing and managing productive land with the oversight and engagement of neighborhood residents, through participatory planning and the recruitment of growers committed to integrating farming into the fabric of their community. Satellites must have some governance role in the central server, however. This helps to ensure that central server staffers keep their focus on supporting neighborhood organizations.

Who Should Own the Land?

In Atlanta and New Orleans, central servers were established so that satellite organizations would hold the land for affordable housing. In an urban agriculture context, it probably makes more sense for a central server to be the landholding entity. As we describe below, this is the approach that NeighborSpace has used to great success in Chicago. It takes advantage of economies of scale and real estate expertise and provides a single point of contact for public agencies that provide land for farms and gardens.

Nevertheless, satellite organizations may want to own the land themselves to better secure local control over neighborhood development. A hybrid approach could provide for initial land ownership by the central server and give the satellite organizations the option to purchase land once they have built their stewardship capacity locally. In the event a satellite organization failed, such an agreement could provide that the land would revert to the central server.

CASE STUDY: NEIGHBORSPACE, CHICAGO

NeighborSpace is the closest existing land trust organization to what might be considered a central server model for urban agriculture. Founded to help conserve Chicago's community gardens, the land trust has recently begun to hold land for urban farms as well. Its history and structure indicate how a central server model might be further developed in other cities. This case study draws upon an article by Ben Helphand (2015), the executive director of NeighborSpace.

The History of NeighborSpace

NeighborSpace was founded in 1996 in response to a recommendation in *CitySpace*, a city planning report that found Chicago lagging behind other major cities in terms of open space per capita (City of Chicago et al. 1998). The report identified the city's fifty-five thousand vacant lots—nearly 15 percent of its land area—as existing open space. Some lots, including community gardens, had already been converted or appropriated for neighborhood use.

CitySpace identified development as a threat to community gardens but noted that no public agency was equipped to own and preserve them. The gardens presented risks and complexities different from parks managed by the Chicago Park District. The report recommended creating NeighborSpace as a land trust to hold urban gardens.

The City of Chicago, the Chicago Park District, and the Forest Preserve District of Cook County joined together to found and provide initial funding for NeighborSpace, which would officially be an independent nonprofit organization (Chicago City Council Committee on Finance 1996). However, the trust operates with support and oversight from its governmental founders. Each founder provides $100,000 per year in funding and holds three seats on the NeighborSpace board. Other seats are filled by staff members of regional open-space land trusts, the University of Chicago, and other nonprofit groups. The NeighborSpace staff has worked to make the organization's benefits apparent to elected officials and parks commissioners, and Helphand has focused increasingly on raising funds from foundations and private donors.

What Does NeighborSpace Do?

As of early 2016, NeighborSpace held just over one hundred sites. Though just a fraction of the hundreds of community gardens in Chicago (Taylor and Lovell 2012), it is nevertheless a sizable amount of land: 23.1 acres of green space. NeighborSpace takes on many of the roles of a central server while leaving certain roles to community organizations. Community gardens are managed by groups of gardeners while the trust holds the title and satisfies insurance requirements (Helphand 2015). Here are some of the functions that NeighborSpace handles:

- **Land and title acquisition.** Much of the land that NeighborSpace holds was donated by the city. Helphand notes that successive administrations and city council members appreciate that the process is predictable. This is helpful because Chicago's powerful city council has broad discretion over transfers of city-owned land in their wards. In one case an alderman required a garden to demonstrate success for three growing seasons before he would approve a land transfer to NeighborSpace.

- **Environmental testing and remediation.** NeighborSpace identifies and addresses legal and environmental risks before taking title to a site, conducting a thorough environmental assessment to avoid taking on cleanup liabilities. When contamination is discovered, NeighborSpace can help community groups secure funding for remediation. These funds have come from open-space impact fees and private foundations.
- **Insurance and tax exemption.** NeighborSpace extends liability insurance coverage to community gardening activities, relieving the gardeners of a major cost. (On urban farming sites held by the trust, the farming organization is responsible for liability insurance.) NeighborSpace also coordinates property tax exemptions for its sites, further reducing land costs for gardens and farms.
- **Water access and stewardship emergencies.** NeighborSpace helps arrange for permanent water hookups at its sites, which can be quite expensive. It promotes water conservation and other sustainable agriculture practices. It can also help community gardeners fix broken infrastructure, manage leadership transitions, and deal with emergencies. For example, a downed tree or someone driving through a fence can threaten a garden's existence, but NeighborSpace's support means that such emergencies can be managed quickly.

There are certain things that the land trust does not do. As much as NeighborSpace helps, it is also careful to leave community organizing to community organizations. Before it will consider securing title to a community garden, for example, NeighborSpace requires a community partner to take responsibility, along with at least three garden leaders and at least ten community stakeholders (Helphand 2015, 2). NeighborSpace also leaves the governance and management of gardens to the community partners, as long as they meet the minimum insurance requirements.

NeighborSpace Expands to Hold Land for Nonprofit Urban Farms

Around 2010 Growing Home, a nonprofit urban farm operating in the south side neighborhood of Englewood, saw an opportunity to expand. Across the street

FIGURE 3. This hoop house on Honore Street Farm allows Growing Home to jump-start its tomato-growing season. (Courtesy of Growing Home)

from a parcel of land it owned—a prior transfer from the city's inventory—was a city-owned vacant lot. Rather than seeking to take ownership of this new property, Growing Home's staff tried something different. It sought to have the parcel transferred from the city to NeighborSpace and then to lease the property from the land trust. This nearly one-acre parcel would become Growing Home's Honore Street Farm (fig. 3).

Up to this point NeighborSpace had held land only for community gardens. Holding land for a commercial farm, albeit a nonprofit one, was a new proposition. The issue prompted discussions by the NeighborSpace board on whether its mission of community-managed open space encompassed urban farming. Ultimately the board agreed that the deal could go forward without an amendment to NeighborSpace's mission or bylaws. In the process it developed basic criteria for holding urban farmland: a farm would be run by a nonprofit organization; it could not be an indoor farm or involve any permanent structures on the site (although hoop houses are permissible); and the site could not be too big. The last criterion remains somewhat vague and depends on the site's context.

City officials, of course, also had to be willing to transfer farmland to Neighbor-Space rather than directly to a farming organization. From the city's perspective, however, ownership by the land trust helps solve some of the problems of site preparation, because the land trust can coordinate and raise funds for environmental testing and any necessary remediation. This can be a significant investment—up to several hundred thousand dollars—and ownership by NeighborSpace helps to ensure that public investment in preparing a parcel is preserved, even if a particular gardening group dissolves or an urban farming organization ceases operations.

The experiment that began at Growing Home's Honore Street Farm has sparked new thinking about how vacant land can be governed and put to productive use. Other projects have followed. In East Garfield Park, a low-income, predominantly African American neighborhood on Chicago's West Side, NeighborSpace now holds 2.6 acres of land that it leases to Chicago FarmWorks, a nonprofit urban farm that grows vegetables for sale at wholesale prices to the Greater Chicago Food Depository (Heartland Alliance 2012). Officials from city agencies and local foundations, eager to expand commercial urban agriculture in Chicago, have come to see the land trust as a useful tool for furthering that goal (Ela in press).

A LAND TRUST FOR FOR-PROFIT URBAN FARMS?

Chicagoans have recently begun to explore how holding urban farmland in trust might also support for-profit, entrepreneurial urban farming models. Foundation officials in particular are interested in helping the urban farming sector move beyond nonprofit, grant-dependent business models. In 2015 a few Chicago foundations joined together to create a program they branded Food:Land:Opportunity, which funded a Neighbor-Space-led effort to develop a land tenure model that could serve for-profit commercial growers in the south side neighborhood of Englewood (Food:Land:Opportunity 2015).

This potential new role for NeighborSpace or a new landholding entity responds to a problem that is likely to arise as a result of the growth of training programs for new commercial urban farmers. In 2013 Mayor Rahm Emanuel announced the creation of Farmers for Chicago, which committed the city to finding land for farmer trainees from organizations such as Growing Home (City of Chicago 2013). The Chicago Botanic Garden (2013) and the Chicago project office of Growing Power (2013) have since developed "incubator farms" through which beginning urban farmers can refine their growing skills, test their business models, and share equipment and distribution facilities.

As new farmers in Chicago's experiment and elsewhere approach the end of their incubation period, it remains an open question where they will go to establish their urban farming businesses. Will they be able to afford land at market rates in the city, or will they have to move beyond Chicago's city limits to find land? Farm incubator programs elsewhere have confronted difficulties in graduating trainees onto their own land outside the program. One of the earliest programs, Intervale Farms in Burlington, Vermont, faced the problem of letting too many early trainees remain on the land as mentors, which meant that eventually little land was left for new trainees (Tursini 2010).

The planning process funded by Food:Land:Opportunity is an effort to help for-profit urban farmers afford land in Englewood. By the end of 2015 the partners had released a prospectus and business plan for Englewood Community Farms, which proposed the creation of a neighborhood community land trust cooperative (Urban Farm Pathways Project 2015). The Ujamaa Community Land Trust is envisioned to initially hold land for farmers but to eventually serve as a locally controlled entity with a mission of shaping housing and business development that is responsive to community interests.

MIGHT THE NEIGHBORSPACE MODEL BE REPLICATED ELSEWHERE?

NeighborSpace is the organization closest to what a central server might look like, but the point here is not that its model could or should simply be transferred to other contexts. The conditions surrounding the founding and funding of NeighborSpace are unique to Chicago. Urban farming organizations elsewhere might need to spearhead a process to create a new land tenure model and seek support from local foundations rather than government partners. That could produce a more formalized network of community-controlled, neighborhood-level satellite organizations than what exists in Chicago. NeighborSpace provides a helpful example, but the structure of new urban farming land trusts will surely vary depending on the contexts and resources available in different cities.

BEST PRACTICES FOR DESIGNING CENTRAL SERVER PROGRAMS

The following list describes the tasks and roles of the central server:

- **Encouragement of government participation.** The vast majority of land for urban farming will probably come from the public sector. In addition, public subsidies for remediation and operations will be necessary in many cases.
- **Engagement with communities.** The central server provides critical support for community-organized urban agriculture but does not dictate how the community must manage land. Community engagement in the governance of central servers is also important for community acceptance, but it may be in tension with government interest in the control of central server functions.
- **Establishment of a clear division of roles and responsibilities.** There should be a clear division of roles and responsibilities among the central server, the government, community organizations, and farmers.
- **Ownership of the land.** Land ownership by the central server may generally work best, but with an option for local entities to purchase (with reversion to the central server if the satellite entity goes under).
- **Performance of the legal and financial work.** The central server should pay particular attention to issues that require technical expertise beyond that of growers and that are best addressed through ongoing relationships between the landholding entity and the government, land banks, and so forth. This expertise applies to title, insurance, land preparation and infrastructure (environmental assessment, remediation, and water), and property taxation.
- **Promotion of communication and education.** The central server can provide for communication and information sharing among growers regarding best practices.
- **Pursuit of opportunities for property tax treatment.** The central server should always seek the most favorable property tax treatment to protect the ongoing affordability of urban farmland. In some cases, this may result in the central server entering long-term ground lease agreements with public entities, trading a bit of control in exchange for property tax exemptions.

CONCLUSION

Farms and gardens are hardly a new feature of America's urban landscapes, having cropped up and withered away repeatedly since the late 1800s (Lawson 2005). This repeated coming and going frames a puzzle: How might urban agriculture become, and remain, a permanent part of our cities? In what ways might we reimagine and restructure land tenure to help urban farmers contribute on a long-term basis to the health (and perhaps wealth) of the cities and communities in which they grow?

Answers to these questions are emerging in fits and starts, as people tinker with ways that urban farmers can gain access to affordable land on a long-term basis. In all likelihood, no single dominant model will emerge, but instead we will see the development of a variety of strategies based on land held by public agencies and by land trusts. In this chapter we have imagined one way in which current models might be extended and expanded to help urban farmers and gardeners become better rooted in their communities. We hope that this model will prove fruitful for grafting onto, hybridizing with, and fertilizing ongoing efforts to reform land tenure rules in cities around the country.

CHAPTER THREE

Growing Urban Food for Urban Communities

ANNE PFEIFFER

In this chapter Anne Pfeiffer presents her observations of the urban farms and gardens she visited in our seven CRFS project cities—the agricultural production component of the food system supply chain—and discusses the significant challenges in simultaneously addressing food production and broader societal goals.

Growing food in urban areas has become increasingly common in recent years. Echoing the pre–World War II past with victory gardens and small commercial farms in cities, personal and community gardens are springing up in containers, backyards, vacant lots, and underused public spaces. Whether driven by an interest in food security or in eating locally grown produce, these gardens and farms face production challenges in a unique environment: space is constrained and soil impacted, unlike in the extensive realms of rural agriculture.

Although the images of urban agriculture generally highlight garden beds bearing a bounty of produce, the goals of urban agriculture projects vary widely and reach well beyond simple food production. Community social goals such as improved food security, job training, and the creation of green space are often central drivers for urban agriculture projects. However, achieving high-quality agricultural production remains important, whether as a goal in itself or as a means to achieve a social goal. A weed-choked vegetable bed with a few cracked tomatoes and a diseased pepper does not serve social, nutritional, or economic goals. This chapter focuses primarily on organizations that seek to produce significant amounts of food on either nonprofit or commercial urban farms, in contrast to community or individual gardens.

IMPROVING ACCESS TO FOOD

Urban agriculture projects can improve food security both directly, through food production, and indirectly, by changing the conditions in which people interact with food, such as by teaching cooking skills, improving economic access, and empowering people and communities. Where access to healthy, culturally appropriate food is limited or nonexistent, urban agriculture can directly improve food security by supplying fresh fruits and vegetables. However, the capacity of urban agriculture to directly improve food security depends on the quantity of food produced, the selection of food items grown, and the manner in which they are sold or distributed.

When evaluating the food security implications of urban agriculture, we must consider the range of crops grown on urban farms and the ability of those crops to meet nutritional and caloric needs. Because of space and policy limitations, urban farms commonly produce vegetables rather than animal products. Among vegetables, crop selection tends toward space-intensive, high-value crops such as greens and fresh vegetables—for example, lettuce, tomatoes, and peppers. The production of storage vegetables and high-protein crops such as beans, winter squash, and root vegetables may be limited because of space constraints or concerns with soil quality. Furthermore, animal protein is typically not produced in urban settings as a result of policy and zoning restrictions. To improve the nutritional dimension of food security, urban growers must intentionally choose crops with high nutritional value (Brown and Jameton 2000).

The manner in which food from urban farms is sold or shared also has a significant effect on the degree to which food security is addressed. To improve food security, as defined in chapter 1, a program must provide affordable, convenient, culturally appropriate food choices. Food produced and sold or distributed in neighborhoods with limited access to fresh nutritious food can be expected to have the largest positive effect on improving food security. On-site farm stands, farmers' markets in underserved areas, market baskets, and community supported agriculture (CSA) programs all have the potential to get local, urban-grown food into the hands of people who need it. An example is ReVision Urban Farm in Boston, which offers a variety of neighborhood programs, including a CSA

FIGURE 4. Growing Power's urban farm at Altgeld Gardens on Chicago's far south side. (Courtesy of Greg Lawless, CRFS)

program, gardening workshops for local residents, a food distribution program at the associated women's shelter, and on-site market stands.

Food must be affordable as well as convenient. Urban farmers must be aware of other sources of food for neighborhood residents and should not expect that low-income residents will choose the locally grown option if it is more expensive than other alternatives, even if the locally grown option is of higher quality. An urban farm created and managed by Growing Power at Chicago Housing Authority's Altgeld Gardens (fig. 4) has struggled to gain customers for its on-site farm stand because the neighborhood residents can acquire less expensive food at a supermarket, despite the fact that it is less convenient for the residents to access.

Urban farms tend to be good at producing culturally appropriate food items: produce that is integral to particular ethnicities, religions, or regions. Foods that are special to the dominant cultural groups in the neighborhoods near an urban farm are often grown to engage and serve the immediate community. Common examples are herbs and cooking greens. The Fondy Food Center in Milwaukee has been particularly successful at providing the supporting infrastructure for farmers of diverse cultural backgrounds to grow and market a wide range of ethnically important crops, and it has also provided accompanying nutrition and cooking demonstrations. The Spring Rose Growers Cooperative near Madison, Wisconsin, has CSA shares that cater to Hmong and Latino clients.

SOCIAL GOALS

Urban agriculture programs often work indirectly toward improved food security through social goals such as job training, skill building, after-school programs, improved health, and increased green space. Youth engagement and education are also common goals of urban agriculture projects. Case studies of individual projects, such as East New York Farms!, indicate that young people engaged in urban agriculture believe that they can develop important interpersonal and job-readiness skills, feel more connected to and invested in their communities, and appreciate the farm as a safe and calm space, which is often absent from other aspects of their lives (Hung 2004).

Several urban farms observed by the CRFS project have a central goal of education and also provide general job readiness and training opportunities ranging from gardening and farming to product development, marketing, and customer service skills. Some examples are an organic farming internship at Community GroundWorks (Madison), construction skills learned through raised-bed and rainwater catchment system construction at Groundwork Somerville (Somerville, Massachusetts), marketing and accounting skills learned while selling produce and flowers with Muir Ranch at John Muir High School (Pasadena, California), retail and food service skills learned at Community Services Unlimited (Los Angeles), and job-readiness skills developed at Growing Home in Chicago (see section). Such education and training programs provide important skills, but they can also provide income to the program participants and improve food security by increasing household income.

Building self-esteem and empowerment is related to educational goals. "It's not about the oranges or butterfly bushes or kale," said Matthew (Mud) Baron of Muir Ranch in Pasadena, a project at a high school whose students are mainly from socially and economically disadvantaged backgrounds. "From my perspective, my kids are broken, they've never finished anything; they've never started anything from a self-esteem standpoint. They don't know how to function in the world. With the plants . . . you put an artichoke in the ground, leave it out in the field,

FIGURE 5. Marquette University's Michael Schlappi and Julie Dawson from the University of Wisconsin–Madison take a tour of Alice's Garden in Milwaukee with its executive director, Venice Williams (center). (Courtesy of Greg Lawless, CRFS)

and with a modest amount of irrigation you're going to have a beautiful nine-foot thistle that tastes good, and the kids are pretty proud of that when they pull it off. And that's how we keep going—because they're proud of it."

In addition to providing job training and skill building, urban agriculture can be a tool for building community. In their study of community garden participants, Teig et al. (2009) found that community garden members had a greater level of mutual trust and social connectedness as a result of their participation. Furthermore, the garden was instrumental in promoting volunteerism, leadership, and civic engagement. Community gardens and urban agriculture can also provide an avenue to teach and strengthen cultural history. Alice's Garden in Milwaukee (fig. 5) offers community garden plots as well as several agriculturally based educational and ministerial programs. A central focus of the garden is to educate and connect people to the cultural history of African Americans as it relates to agriculture and food.

Urban agriculture can also provide a space for immigrants who have recently left more agrarian landscapes and lifestyles to preserve a piece of their culture and livelihood. The Fondy Food Center operates a farm north of Milwaukee where farmers can rent plots for commercial production. Many of these farmers are Hmong with strong agricultural backgrounds, and they view their farmwork as a vital connection to their traditional culture.

In evaluating urban agriculture, we must be mindful that cultural connection and personal fulfillment are central to many farms and gardens and cannot be quantified in terms of a typical profit-and-loss statement. Cost-benefit analysis, discussed in chapter 14, provides information about the value of such initiatives. Urban agriculture can be an effective tool in achieving community improvement goals, but it is critical to realistically balance these goals with reasonable production targets.

Growing Home, Chicago

BY APRIL HARRINGTON, HARRY RHODES, AND REBEKAH SILVERMAN

As Chicago's first and only USDA-certified organic production farm, Growing Home focuses on job training and employment in the context of an urban farming business. Through our work in Englewood on Chicago's south side, we attempt to fight chronic unemployment, lack of food choices, and uprooted families through living wages, access to healthy food, and stable households for families in the community.

Growing Home's vision is a world of healthy people and communities. Our mission is to operate, promote, and demonstrate organic agriculture as a vehicle for job training, employment, and community development. Les Brown, Growing Home's founder, described the philosophy behind our work. "People in our program are often without roots," he said. "They're not tied down, not connected, not part of their family anymore. Our organic farming program is a way for them to connect with nature—to plant and nurture roots over a period of time. When you get involved in taking responsibility for caring for something, creating an environment that produces growth, then it helps you build self-esteem and feel more connected." Today we work with people who face multiple barriers to employment in addition to housing instability.

Growing Home has been using organic agriculture to change lives since 2002. Through hands-on training on our high-production, organic urban farms, our motivated production assistants not only gain proficiency in organic farming, they also learn techniques to demonstrate their positive attitudes, reliability, and commitment and to improve their speed, efficiency, and problem-solving abilities.

We have been working in Englewood for more than ten years. In our community, food choices and economic opportunity are limited. Our farming business brings organic produce and jobs to our program participants and graduates, their families, and our neighbors. We were the first farm in this community and the catalyst for a future urban agriculture district. With our partners we are using farming and food enterprises to transform Englewood from a "food desert" into a food destination. Today Growing Home operates two urban farms in Englewood and is planning its third. The neighborhood is home to two additional farms operated by other organizations, many community gardens, and two independent cafés; under construction is a Whole Foods Market, a Chipotle Mexican Grill, and a Starbucks. Together we are creating hundreds of jobs and changing this community.

We sell USDA-certified organic produce at reduced prices to our neighbors in Englewood, and we have increased sales at our on-site farm stand by 600 percent since 2011. We pair fresh vegetables with outreach and education through cooking demonstrations, garden workshops, open houses, and tours. In 2015 we harvested more than thirty thousand pounds of fresh organic produce on less than one acre of land.

In 2015 we trained forty people who have struggled to find or keep employment because of criminal records, homelessness, substance abuse, lack of education and job experience, or other barriers. In 2016 we trained fifty people, and in 2017 we will train sixty. Although many job training programs exist in Chicago and across the nation, only a handful pay participants for every hour they work and study, as we do during the fourteen-week program.

We work to prepare our graduates for full-time jobs throughout Chicago's food chain—from growing to packaging to preparing and selling food. All our production assistants take a certification course for in-house ServSafe food handling and learn transferrable skills on our farm. When they acquire jobs, they typically earn above minimum wage and enjoy benefits and promotions. In 2014 and 2015 eighty people participated in our transitional jobs program. During that period, 80 percent completed the program, and 80 percent of the graduates found full-time employment. Many of them continue to volunteer through our graduate board so they can give back to society and serve as mentors for people with similar backgrounds.

Growing Home's unique contribution to urban agriculture is its commitment to our social mission. We operate a successful, highly intensive organic urban farm. With five hoop houses, our farm serves as an example of year-round growing in Chicago. We use our farm to change lives for individuals and communities.

MAKING URBAN AGRICULTURE WORK

Many aspects of urban and rural agriculture are similar, but some production techniques and many aspects of business management are quite different for urban growers than for their rural counterparts. Production management strategies are shaped by urban land access and tenure arrangements, space limitations, crops or other items produced, the background and experience of labor, infrastructure, business models, financing, and policy. Urban growers are using a variety of innovative approaches to respond to the unique conditions of urban areas. The rest of this chapter examines urban agricultural production through the lens of practical considerations for people who build urban agriculture projects.

Space-Intensive Production

Urban farms operate almost universally with limited access to land. Even where vacant urban land is plentiful, parcels are generally very small compared to the size of rural farms. The limited space of urban farms and their proximity to large populations shape their production. Because space is at a premium, close plant spacing techniques are ubiquitous on urban farms. Careful crop selection is critical for farms that operate with limited space. Sprawling crops, such as vining squashes and melons, may be difficult to include in an intensive production system. Similarly, the growing days required for a particular crop determine the degree to which it can pair well with other crops in an intensive rotation. Some crop varieties are better adapted to close spacing. Varieties that bear continuously once the plants are mature may be preferred to plants that ripen all at once (determinate varieties).

Livestock may be successfully integrated into a space-intensive production system. Vermicomposting (the use of earthworms), aquaculture, and apiculture (beekeeping) are the most common examples; chickens, ducks, and goats are common where zoning allows. The inclusion of livestock on urban farms contributes protein, as well as fertilizer in the form of manure, to urban food systems.

The use of growing techniques such as season extension and double-cropping, which increase the effective amount of available land by growing multiple crops in the same area, can also increase yields. Growing Power in Milwaukee uses a double-cropping technique to ensure nearly continuous production by overseeding subsequent crops of greens into beds that are still producing. These beds consist entirely of nutrient-rich vermicompost and coir (shredded coconut husk) mixed by Growing Power, thus increasing the production per unit area.

Other season extension techniques, such as hoop houses, are common to both rural and urban agriculture but are used in urban areas to maximize the production return per land area. Greenhouses, hoop houses, and high tunnels allow production of high-value crops for longer portions of the year, which improves overall food output and increases the time span in which nutritious food can be provided to community members. In a climate where the frost-free period is May to September, Growing Power can produce spinach, kale, and other cold-hardy crops year-round in its hoop houses by capturing heat from compost generation. In addition to increasing food production, these season extension techniques can be important in continuing community involvement beyond the summer growing season. Growing Power's greenhouses and The Food Project's Dudley Street Greenhouse in Boston act as sites for cold-weather community gatherings and educational events.

Nontraditional Spaces

Urban growers often access land by operating in nontraditional spaces. Repurposing warehouses, rooftops, and parking lots allows growers to find space where land might not be available. This can facilitate urban agriculture in densely populated areas, close to the community and the potential markets the project wishes to engage. Furthermore, these nontraditional spaces create unique growing environments that facilitate the production of specialty crops. Growing Power's Iron Street Farm in Chicago occupies a former trucking warehouse and is able to grow greens and mushrooms, raise fish, and produce vermicompost indoors, along with growing other vegetables in outdoor hoop houses and raised beds constructed atop a former parking lot. Products that benefit from a controlled climate, such as fish, and those that can easily be grown under artificial or limited light, such as microgreens and mushrooms, are a good fit for these environments. Many of these foods have the added benefit of being high in value. Several blocks from Iron Street Farm, The Plant, located in a former meatpacking facility, has capitalized on the many unique features of indoor food production and acts as an incubator for nascent urban agriculture businesses.

Producers who use nontraditional locations should be aware of possible production challenges unique to their spaces. For example, rooftop gardens are subject to high winds and high evapotranspiration (loss of water from plants and soil to the atmosphere), so growers need to create windbreaks and choose crops or varieties that are especially tolerant to such conditions. Rooftop gardens must also be designed with a consideration of the structural load that the building can support. Lightweight growing media, such as coir (fiber from coconut husks) or specially manufactured rooftop growing mats, may allow rooftop gardens on a larger range of buildings. On the ground, farms on top of asphalt require strategies for dealing with the lack of drainage and more dramatic temperature fluctuations. In dry climates, lack of water is a concern for these nontraditional gardens; when plants use more water, costs can increase.

Rural-Urban Linkage

Given the challenges of producing food in the limited space of an urban environment, creating a linkage with rural farms can provide ways for urban growers to expand their production capacity while maintaining the community development benefits of growing in an urban area. Operating an urban farm with a nearby rural counterpart allows the urban parcel to serve the community with convenient public access to the urban farm for activities such as volunteer involvement, after-school programs, job training, and other community-centered programs. Being located near a population center may also assist urban farms in achieving the goal of delivering fresh food to communities underserved by grocery stores. A rural linkage allows for a larger-scale production of crops, perhaps using more mechanized techniques. This can facilitate economies of scale that result in less expensive food, space-intensive crops that are impractical on confined urban parcels, and livestock that may be prohibited in urban areas. Many of the urban farms included in the CRFS project had a counterpart urban fringe or rural site.

Value-Added and High-Value Crops

In addition to utilizing space-intensive growing methods that maximize the amount of food grown, urban farms can optimize space use by producing high-value crops or crops that can be converted into value-added items. This strategy is very common among commercial urban farms as well as with noncommercial and hybrid models that seek to boost revenue. Urban farms can improve profitability by selling high-value crops and value-added products in high-end markets. These crops include microgreens, flowers, fish, mushrooms, and honey. Urban producers can add value by pickling, fermentation (e.g., kimchee and sauerkraut), and jellying. However, such products are not likely to improve food security (Brown and Jameton 2000). The pressure to produce a return on highly valued land in urban areas is likely to be at odds with a food security goal, although some organizations balance social goals with income needs by selling a portion of their products to high-end markets while ensuring that the other portion is made available to neighborhood residents. Growing Home and Growing Power in Chicago both sell at the upscale Green City Market farmers' market in Chicago and achieve their social goals through their other programs. For some organizations, selling produce at a premium price can teach important marketing and business skills, which may transfer indirectly to improved food security by providing tangible job skills.

Labor and Management

The labor and management of urban farms tend to be very different from those of traditional rural farms. Urban farms with a social mission are often managed by a staffer who was hired into a more general position without significant agricultural experience. In these cases, improving the production skills of farm managers can increase yields. Commercial urban farm managers are more likely to have strong agricultural backgrounds, but they still face limited resources. Whereas rural growers often rely heavily on peer groups, consultants, and university extension departments for information exchange and support, urban growers often lack a robust network of experienced, knowledgeable growers. Furthermore, urban extension programs are no longer common, and urban agricultural research stations are almost nonexistent. This limits the availability of accessible professional assistance to urban farmers. Professional development and a greater availability of resources can increase the knowledge and skill of urban farmers, resulting in increased yield and higher-quality production.

Urban farms with community engagement goals are very likely to use volunteer labor. Space-intensive production techniques often require a high degree of manual labor, which may be a good fit for urban farms with plentiful community volunteers eager to be involved in farmwork. However, volunteers often have limited agricultural experience and may not be able to perform all farm tasks reliably. Also, managing the volunteers can redirect staff time away from farming. The experience level of the farm labor, including volunteers, must be considered when selecting crops. Farms that rely heavily on volunteers may benefit from avoiding crops with more challenging maintenance and harvest needs. Supervising young volunteers has the additional challenge of managing energetic but easily distracted participants. This challenge is balanced, however, by the satisfaction of making a difference in their life trajectories.

Infrastructure

Urban farms face a number of infrastructure challenges. The ability to construct necessary infrastructure, such as hoop houses and other structures (e.g., sheds for tool storage or for produce washing and packing), may be limited by city policy and zoning, the lack of available financing, and unstable land tenure. In addition to requiring field space, farms need facilities for growing seedlings under shelter, places to store the tools and equipment necessary for production and harvesting, and places to wash, pack, and store produce.

For many crops, seedlings started in a hoop house or a greenhouse and then transplanted will generate greater yields than seeds sown directly into the soil. This will be a disadvantage for a farm that doesn't have a seed-starting facility and can't develop a collaborative relationship with another entity that does. Limited storage or packing facilities (including cold storage) may limit the crops that can be grown or the marketing channels available to the farm. An operation that lacks cold storage, for example, will need to harvest all produce shortly before market, limiting its overall growth potential, and may choose not to grow crops that require quick chilling for optimal quality. The lack of a packing shed may also limit a farm's ability to comply with many food safety practices.

Urban farms may be susceptible to vandalism. An urban farm in Detroit once had an entire hoop house stolen, presumably for scrap metal. Community GroundWorks in Madison had its farm truck and tractor vandalized repeatedly. Community engagement strategies can be helpful in minimizing such instances by relying on neighbors and other community allies to keep watch and foster a culture of support for the urban farm. Community GroundWorks has included a "pick your own" component in its CSA project as a way to increase public traffic to the relatively secluded portion of the farm that was vandalized.

Economics

Urban farms operate under a variety of business models. The particular model chosen by a farm influences a broad range of production decisions and priorities. As businesses, urban farms fall on a spectrum between social mission–driven projects and entrepreneurial commercial farms, with many examples of hybrid mission–driven businesses in between. On one end of the spectrum, mission-driven organizations tend to use farming to achieve a social goal such as neighborhood engagement, youth education, job training, or green-space creation. Food production may indeed be a goal, but it is generally shared or secondary to some other goal.

In such cases, production decisions will focus on the primary goal. For example, a farm project that seeks to engage neighborhood residents and schoolchildren should select crops that can be easily harvested and cared for by both groups. The physical layout of the farm should accommodate the ability of people to move easily through rows and provide a gathering space for community members. Alice's Garden in Milwaukee provides a good example of this design with its historical and cultural demonstration gardens, living labyrinth, and covered picnic shelter where community members gather and share meals.

In contrast, organizations with a primary goal of producing for food-insecure communities may more closely resemble a commercial farm. For a farm to effectively produce as much food as possible on a plot of land, the management must have agricultural experience, and production decisions should be made with the goals of high yield and quality. Some volunteer labor may be used to keep production costs down, but the operation is most effective when volunteers have agricultural experience or can commit to working on the farm regularly.

The Hunger Task Force Farm outside Milwaukee is an example of this model. All of the food produced is distributed through local food pantries. The two-hundred-acre farm has a professional farm manager and staff while relying significantly on volunteer assistance. Food is produced on a large scale, using row-crop techniques and mechanized equipment whenever possible. Notably, the Hunger Task Force Farm follows the rural-urban linkage model of capitalizing on a large parcel of less expensive rural land to grow large volumes of food for urban populations in need.

Alternatively, a for-profit commercial farm in an urban area is likely to focus on maximizing high-value production and creating intensive rotations that support multiple crops each year. Crop selection in this case will focus primarily on the value of crops rather than serving community or social goals, with labor and management provided by people with skilled agricultural backgrounds. Crops are often sold at urban farmers' markets, to restaurants, or to CSA customers.

Of course, many hybrid models fall between the ends of the spectrum and attempt to support both commercial food production and the achievement of social goals. The successful implementation of a hybrid model requires a clear vision and mission, since the organization must be able to prioritize the conflicting needs of the social and commercial sides of a hybrid farm project. Growing Power's operations in Milwaukee and Chicago are prime examples. The organization produces high-value crops, such as microgreens and mushrooms, that are sold to upscale restaurants and are farmed by highly skilled agricultural workers. Growing Power simultaneously supports a number of neighborhood-based garden or farm sites and a job-training program.

"That's what we're all about, putting people to work," said Will Allen of Growing Power. "Sure, I could bring a tractor out here with a planter, and I'd have this whole thing planted in a couple of days. But where would that put us? The whole idea, our mission, is to empower people and put them to work."

Regardless of the business model chosen, production methods that maximize the quality and quantity of food produced can improve the financial viability and food output of an urban farm operation, whatever its primary goal may be.

Agriculture and Food Policy

Local regulatory policies such as zoning play a critical role in shaping urban agriculture. Policies that overtly support urban agriculture can be pivotal in creating a successful, dynamic community of growers, whereas restrictions, prohibitions, or vague policy language can make the urban food environment unsupportive or unpredictable (see chapter 12). Urban growers often operate under policies that were designed with home landscaping in mind, with little appropriateness to agriculture. Examples are limits on plant height and requirements that front lawns must include a certain percentage of grass.

Common policies supportive of urban agriculture include provisions that allow the keeping of bees or chickens or that allow temporary farm stands. Although advocates often press for policies that explicitly allow urban agriculture practices, these policies may end up creating additional restrictions. In 2013 the city of Boston passed Article 89, which supports the expansion of urban agriculture but requires that urban farms conduct regular testing of their soil, water, and soil amendments such as compost and other fertilizers (Boston Redevelopment Authority 2013).

This testing far exceeds the requirements for rural farms and adds to the costs of urban farming.

In Cedar Rapids, Iowa, city policy was written to explicitly support urban agriculture as an economic development strategy in floodplains after the devastating floods of 2008 (Cedar Rapids, Iowa 2012). However, this policy limits the equipment used in production to nonmechanized tools and prohibits the use of any input not intended for home garden use. Although the spirit of the policy is supportive of urban agriculture, the statute in effect creates additional restrictions and a regulatory landscape in which financially viable, ecologically sound farming is all but impossible.

CONCLUSION

Urban agriculture is multifunctional (Lovell 2010), with many approaches dedicated to meeting various goals. In general, most urban farms balance three broad goals: growing food, building or serving community, and making a profit or earning a livelihood. Urban growers are capable of producing significant quantities of high-quality food when they have the experience and resources necessary to make sound management decisions. However, high-yield production is often not the primary goal of an urban agriculture operation. In such cases success must be measured by more complex, nuanced metrics that account for the myriad community and social benefits achieved by a particular agricultural program. Furthermore, individual organizations must strategically set goals and prioritize activities that meet those goals. While recognizing that food production may not be the highest priority, any urban farming organization must examine the role of food production in its program and evaluate the extent to which the food's quality will bolster the success of the programs that food production is intended to support.

CHAPTER FOUR

Distribution

Supplying Good Food to Cities

LINDSEY DAY-FARNSWORTH

In this chapter Lindsey Day-Farnsworth presents many challenges, at different scales, in aggregating and distributing local food production through values-based supply chains and distribution strategies, and she introduces some promising initiatives. This chapter represents the distribution and aggregation component of food system supply chains.

In the past century, high-volume conventional food distribution systems have achieved remarkable efficiencies. Yet these gains have not come without a cost. Many conventional food supply chains present challenges to farmers and consumers alike. For farmers the central challenges are market access and profitability. For consumers the issue is accessibility, especially in low-income communities that lack adequate healthy food retail options. This chapter explores innovations in community and regional food distribution that address these challenges through a diversity of strategies, such as producer-oriented capacity development, new approaches to supply chain governance, healthy food retail initiatives, and non-traditional farm-to-market programs. Examples of specific innovations used by CRFS project partners are woven throughout the chapter.

BACKGROUND

Despite the productivity and efficiency of conventional food systems in the United States, their environmental and social costs are well documented. Characterized by large-scale, input-intensive, monocultural production and long-distance transport, the conventional food system contributes to environmental problems, such as natural resource contamination and depletion, decreased biodiversity, and climate change. Decisions about crop selection, production practices, and food distribution have a significant effect on genetic diversity and greenhouse gas emissions. Consider, for example, that we have lost as much as 75 percent of the genetic diversity of agricultural crops worldwide (Thomas et al. 2004), making them vulnerable to catastrophic loss from pest and pathogen epidemics, and that food system emissions contribute 25 to 30 percent of total global greenhouse gas emissions, depending on assumptions about what can be attributed to food production and distribution (Consultative Group on International Agricultural Research n.d.).

The social issues that characterize the conventional food system are similarly complex and far-reaching; they include unfair distribution of risks and profits across food supply chains as well as racial disparities in nutrition, access to healthy food, and labor and compensation practices. Recent estimates suggest that farmers capture just over ten cents of the average food dollar, which is about half the value captured by the food processing industry and less than one-third of the share captured by the food service industry (Canning 2011). Farmers are not the only people being squeezed; many food chain workers—from field laborers to short-order cooks—work in substandard conditions and struggle to make ends meet (Jayaraman and Schlosser 2013).

Research also shows that people of color make up a disproportionately high percentage of the lowest-paid positions throughout the food industry (Liu and Apollon 2011). This racial divide not only characterizes labor and compensation patterns *within* the agrifood system, it also extends to food insecurity and public health statistics and to the geography of food access. Data from the USDA Economic Research Service (2012) indicates that 14 percent of the US population experienced food insecurity in 2012; about one in seven people did not have sufficient food at times, and black and Latino households experienced greater food insecurity than other racial and ethnic groups. Female-headed households are more likely to experience poverty and food insecurity and have higher rates of obesity (Martin and Ferris 2007). Similarly, there is a disproportionately high prevalence of obesity and diet-related diseases in black and Latino populations in the United States (Cossrow and Falkner 2004; Kumanyika and Grier 2006).

Community and regional food distribution systems are a direct response to conventional profit- and volume-driven food distribution systems. As such, they are often driven by values such as justice and fairness, strong communities, and vibrant local economies (Abi-Nader et al. 2009). Although local and regional food distribution systems are not inherently more socially just or ecologically sustainable than conventional food distribution systems (Born and Purcell 2006), their scale and place-based nature have made them ideal for experimenting with ways to explicitly integrate values into food supply chains. Community and regional food systems are diversifying food distribution channels by developing new approaches to activities such as sourcing, aggregation (the consolidation of products from multiple farmers to achieve the volume necessary for large retail and institutional markets), processing, supply chain governance, marketing, transportation, and consumer access (Day-Farnsworth et al. 2009).

CHALLENGES FACING COMMUNITY AND REGIONAL FOOD DISTRIBUTION

Even when proponents of community and regional food systems share a broad vision of an economically vibrant, equitable, and sustainable food system, the strategies they use to fulfill this vision are likely to differ depending on their personal and professional backgrounds. For example, producers and advocates whose backgrounds are in rural sustainable agriculture typically focus on farmers' livelihoods and the environmental and health effects of production practices, whereas urban food activists predominantly focus on issues of food access and household food security. Although these goals are not contradictory, the challenge of advancing them simultaneously evinces the divergent stories in the history of the good food movement and underscores a central tension within the ideal of community food security (Gottlieb and Joshi 2010). This tension is perhaps most pronounced in the configuration of food supply chains, because the challenges that characterize community and regional food distribution tend to fall into one of two categories: access to economically viable markets or access to healthy food. Figure 6 shows where these challenges manifest in community and regional food distribution systems.

Correspondingly, producer-oriented distribution strategies are principally concerned with the question "How can we reduce barriers to market and configure supply chains so that small and midsize local and sustainable producers can make a living?" Meanwhile, strategies oriented toward food security address the question "How can we reconfigure food supply chains to ensure that everyone has access to fresh healthy food?" This chapter first explores strategies that have been used

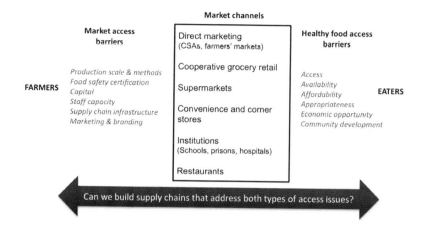

FIGURE 6. Barriers to market access and healthy food access in food supply chains. (Courtesy of Lindsey Day-Farnsworth, CRFS)

to address these challenges separately. It then showcases several examples from CRFS research partners that strive to address both issues at once.

POTENTIAL ADVANTAGES OF REGIONAL VALUES-BASED FOOD SUPPLY CHAINS

Although many people are familiar with direct marketing channels such as farmers' markets, farm stands, and CSA programs, *intermediated* (or multistage) food supply chains are less visible components of community and regional food systems. Despite their growth and high visibility, direct marketing channels actually make up a relatively small percentage of total local food sales. According to the USDA, farms that sell through intermediated channels reported $2.7 billion in local food sales in 2008, more than three times higher than the value of local foods marketed exclusively through direct channels (Low and Vogel 2011). This trend is not likely to change anytime soon.

Direct marketing is unlikely to supplant other distribution channels for two primary reasons. First, market research suggests that consumers seek local food through a variety of market channels, including grocery stores, restaurants, and institutions such as schools and hospitals. In fact, not only do most US residents purchase the majority of their food from large grocery retailers (Hartman Group 2008), more than 85 percent of consumers report that the availability of local food is an important factor in deciding where to shop for groceries (National Grocers Association 2012). Similarly, more than 60 percent of US adults report that locally sourced menu items are important to them when choosing a full-service or quick-service restaurant (Miller and Washington 2014). As a result, retail grocery stores, restaurants, and institutions are increasingly seeking locally and regionally sourced products, and often in volumes that exceed the production capacity of small-scale local farmers.

Second, direct marketing is not a good fit for many producers and products. Direct marketing channels excel at moving low volumes of high-quality products at high prices per unit. Consequently, they are best suited for small-scale farmers, such as diversified vegetable growers, who produce small volumes of numerous fresh market crops. By bypassing distributors, these producers can ask for higher prices than wholesale producers and can capture the entire value of the retail food dollar. However, for medium-scale producers, who typically grow higher volumes of fewer products—for example, they grow more than they can sell through farmers' markets and CSA groups but not enough to compete in commodity markets—it is impractical to sell only several pounds of a product at a time to household consumers. The problem for medium-scale producers is that their output is often too small to achieve the economies of scale necessary to compete in high-volume commodity markets (Stevenson and Pirog 2008). Small- and medium-scale producers also miss out on access to processing facilities that use products consumers will not buy directly—those that are blemished, undersize, and in other ways imperfect.

Intermediated regional food supply chains can link medium-scale farmers with area markets, thus filling the demand for higher volumes of local products while offering higher unit prices than commodity markets. Yet as product volumes increase and supply chains are lengthened, new strategies must be developed to maintain the social, ecological, and economic values that have become associated with direct-marketed local food. In the past fifteen years researchers have identified the key characteristics of particular intermediated food supply chains that promote a combination of social and ecological benefits while ensuring a fair economic return for small- and medium-scale producers.

Some defining features of values-based food supply chains include the explicit use of social and ecological values, in addition to profit and quality considerations, to guide business decisions; a commitment to equitable relationships between supply chain partners; and marketing and branding practices that differentiate values-based food supply chain products from commodity products by highlighting product origins and production characteristics. For example, values-based food supply chains usually have a high level of cooperation between supply chain partners in setting prices, and they often use sustainable production practices, such as the USDA organic standards. Finally, these unique business networks typically supply retailers and restaurants rather than direct markets.

Characteristics associated with product origin or production practices are called *credence attributes* because they are not as readily identifiable to customers as product characteristics like price or quality (Pullman and Wu 2012). Although some consumers are willing to spend more on products with credence attributes, these attributes must be communicated in order to capture a premium—that is, a higher retail price. Products with credence attributes are typically differentiated in the retail environment through marketing claims made on ecolabels that verify particular production practices (e.g., humanely raised, grass-fed, or organic) or through packaging and merchandising that link products to particular farms or regions. Values-based supply chains advance socioeconomic and ecological values through the strategic use of marketing to promote credence attributes in addition to employing innovative approaches to supply chain governance and logistics management.

Increased Market Access through Capacity Building and Infrastructure Development

Premiums associated with sustainably produced foods help internalize social and environmental costs, enabling retail food prices to more accurately reflect the cost of production. Yet to successfully occupy this market niche, producers must adhere to wholesale production and postharvest handling standards and develop appropriate marketing and branding strategies. Research suggests that producers who move from direct marketing into differentiated wholesale regional markets encounter many challenges as they increase production volume. These challenges include difficulties controlling product quality and consistency, overcoming short growing seasons, matching supply and demand, marketing and branding the products, accessing capital and appropriate supply chain infrastructure, developing capacity, and maintaining information flow and transparency across the supply chain. Information is needed so consumers have greater knowledge of the origin and production practices behind the food they eat and farmers have greater bargaining power with their supply chain partners (Day-Farnsworth et al. 2009).

Increasingly, distributors and other supply chain partners are recognizing the importance of helping producers overcome these barriers in order to ensure product quality and marketability. As a result, it is not uncommon for supply chain partners to provide technical assistance, infrastructure, and other support for their strategic suppliers (Stevenson and Pirog 2008). For example, local processors or distributors of food might provide assistance with varietal selection; training in food safety, packaging, or marketing; and product aggregation and branding.

Social and Economic Benefits for Farmers through Equitable Supply Chain Governance

Values-based supply chains are characterized by high levels of transparency and collaboration between supply chain partners, which often includes an explicit commitment to "equitable power relations" and the welfare of all supply chain participants (Lerman 2012). Stevenson and Pirog (2008) emphasize that trust, information flow, and fair governance practices are mutually reinforcing in strategic supply chain partnerships. They note that while supply chain partners do not need to have equal decision-making power, successful values-based supply chain governance requires a commitment to both procedural and distributive fairness. The former ensures that policies and practices are in place so that less powerful supply chain partners (often farmers) can negotiate with more powerful partners in the chain (typically processors and retailers), whereas the latter ensures that risks and profits will be distributed across the supply chain.

For example, the Wisconsin-based cooperative Organic Valley has a mechanism for redistributing unexpected earnings to the suppliers of its produce program and ensures that preseason price negotiations serve as a price floor for growers rather than a ceiling. Organic Valley orchestrates production planning, sales and marketing, and logistics for its produce growers. During the growing season it pays its producers biweekly according to prices negotiated in advance of the season. However, if the annual revenue exceeds expectations, the cooperative pays its growers a year-end "pooling bonus" based on total revenues minus the base price and cost of freight and commission (Day-Farnsworth et al. 2009).

Reduction of the Environmental Impact of Food
Supply Chains through Regionalization

Food miles have become a common proxy for the negative environmental impact of food supply chains. However, the carbon footprint of the transportation segment of food supply chains is a function of much more than miles traveled; it also depends on factors such as load size, vehicle type, and logistics practices (King et al. 2010). In fact, King and his colleagues found that with the right equipment, higher-volume regional supply chains can maximize vehicle efficiency and load size more easily than most small-scale truck farmers while minimizing the distance traveled. This means that high-volume regional food supply chains have the potential to be more resource efficient with fuel and greenhouse gas emissions than ad hoc local distribution.

The growing volume of road freight also has labor, health, and safety implications. Viscelli (n.d.) notes that improvements to rural-urban freight transition have the potential to improve "highway safety, public health, CO_2 emissions, air pollution, traffic congestion, road and infrastructure damage, driver productivity, health and retention, and logistics costs." Hauling equipment upgrades and improvements in logistics are largely the purview of the private sector, but public and nonprofit entities can encourage haulers and distributors to use more size-appropriate, fuel-efficient vehicles and more strategic routes by developing and maintaining transportation infrastructure to promote environmentally efficient distribution systems.

The section "Regional Food Freight" showcases promising models for ways in which public and nonprofit investment in peri-urban freight ports, food terminals, and other forms of distribution and aggregation infrastructure could help farmers move more fresh products into large metropolitan markets more efficiently. These innovations could play an important role in helping regional food supply chains reduce their carbon footprint while making regionally sourced food more accessible to households at all income levels.

Through a combination of strategic marketing and branding, coordinated technical support and infrastructure development, and innovations in supply chain governance and logistics, values-based local and regional supply chains can promote sustainable production practices, support farmers' livelihoods, and reduce the carbon footprint of our food distribution. However, the premiums that make these supply chains economically viable for producers can place a burden on low- and middle-income households. The next section explores the other side of the challenge for community and regional food distribution.

Regional Food Freight

MICHELLE MILLER

A 2015 USDA report indicated that while direct market sales of local food plateaued between 2007 and 2012, intermediated sales—mid- to high-volume sales to wholesale buyers—continued to increase (Low et al. 2015). This growing marketing channel is compromised by an underdeveloped infrastructure for regional food freight. Unlike national food supply chains, whose infrastructure is mostly privately owned, regional food supply chains are not vertically integrated and often lack access to the distribution and logistics infrastructure that is necessary to reach markets efficiently. This is an important aspect of diversifying production and matching it to market scale. Regional food economies would benefit from two types of distribution infrastructure: terminal markets and truck ports.

Terminal markets are important for mid- and small-scale metropolitan food distribution systems because they can collect and distribute regional, national, and international food products. The Ontario Food Terminal in Toronto is a successful example of a nonprofit food terminal with a public mission to support Canadian supply chains. It provides critical storage and cross-docking infrastructure (i.e., facilities where semitrailers directly transfer food products to other delivery vehicles) for wholesale growers, distributors, and buyers of all sizes.

The Ontario Food Terminal is the third-largest food terminal in North America; it provides cross-dock space for about twenty larger distributors and serves about four hundred farmers selling wholesale from temporary stalls. Nearly five thousand wholesale buyers are registered to do business, from buyers at larger chain stores to independent caterers. Growers from a two-hundred-mile radius bring products to the terminal, and buyers come from much farther away. The Ontario Food Terminal is an anchor of the regional food economy, with an estimated one hundred thousand direct and indirect jobs attributed to it in the Great Lakes region (Lengnick, Miller, and Marten 2015). This example has much to teach US food system planners; similar infrastructure exists in cities such as Chicago and Boston and could be adapted to serve a wider range of growers, buyers, and distribution needs.

Truck ports and related peri-urban freight infrastructure are necessary to help growers and haulers serve urban markets more efficiently. Urbanization and population growth are national trends. By 2050 much of the US population will probably reside within eleven megaregions, further increasing traffic congestion in metropolitan areas. Congestion dramatically reduces freight fuel efficiencies, increases labor and logistics costs, and results in high greenhouse gas emissions. Recent design innovations have improved freight transportation efficiencies both for long-haul trucks (those that move products 150 miles or more) and smaller urban freight vehicles. By specializing freight movements between the rural and urban segments of a trip, the food transportation sector could take advantage of these innovations to improve fuel and labor efficiencies and reduce greenhouse gas emissions. However, these improvements require infrastructure specific to large urban centers.

Peri-urban truck ports are freight container transfer stations outside metropolitan traffic congestion areas that serve as exchange yards where distributors can align load and vehicle size, drop off or pick up trailers to make urban deliveries during off-peak hours, and charge electric delivery vehicles (Viscelli n.d.). At a peri-urban truck port, a trucking company could exchange the truck tractor optimized for rural long-distance driving for a tractor optimized for urban delivery, including electric or alternative-fuel vehicles. No product is unloaded—the containers are simply switched to tractors specific to urban or rural driving conditions. The trucking company could also compensate drivers differently, for their time in urban conditions and for miles driven on long-distance segments.

The University of Wisconsin–Madison Center for Integrated Agricultural Systems (CIAS) found that growers who want to supply metropolitan Chicago cannot efficiently ship products into the region because of congestion and other challenges associated with current supply chain configurations. To address this issue, CIAS researched options for improving regional growers' access to Chicago markets. Despite concerns about supply chain governance and trailer ownership, growers and carriers (businesses that transport products) expressed interest in truck ports because of their potential to facilitate logistics in metropolitan regions, especially as new regulations limiting driver hours go into effect. Although accessible truck ports will require public and private investment and coordination, the private sector already offers examples of how these facilities can promote more efficient metropolitan food distribution.

For instance, C. R. England (2015), one of the largest refrigerated carriers in the United States, was honored by the US Environmental Protection Agency (EPA) Smart-

Way program for its high-quality environmental performance. The company, which operates four terminals in four states (California, New Jersey, Indiana, and Texas), recently invested in a truck port fifty-six miles outside Los Angeles. As a dedicated contract carrier—a trucking company contracting with a specific shipper to move products along regular routes—it is relatively straightforward to swap a truck tractor designed for long-distance hauling with another tractor for the urban segment of a trip. This practice has allowed the company to shift some of its fleet to alternative fuels, adopt long-haul efficiencies for other tractors, and improve overall logistics.

Although these practices are probably easier to implement within a single company, the development of public or nonprofit truck port infrastructure could significantly improve independent growers' access to metropolitan markets by improving freight flow into cities. Smaller, regional trucking companies may benefit immediately from upgrading their fleet with urban optimized trucks, especially if regular routes can be established between aggregation facilities and markets. The industry might also benefit by extending fleet improvement incentive programs, like those offered by Chicago Area Clean Cities, beyond metropolitan areas to rural companies that move food regionally.

RETHINKING EQUITABLE FOOD DISTRIBUTION

According to the USDA Economic Research Service (2013), approximately 14 percent of US households are food-insecure. Although a household's economic constraints are a major factor in household food security, related issues, such as a lack of healthy neighborhood food retail outlets and limited transportation, exacerbate the problem in many low-income communities. Nearly 10 percent of the US population lives in low-income areas more than one mile from a supermarket (Ver Ploeg et al. 2012). Furthermore, almost one-third of the US population does not have access to a car or to reliable public transportation (PolicyLink, Reinvestment Fund, and Food Trust n.d.). Those who *can* take public transportation to full-service supermarkets are limited in the amount of groceries they can carry on one trip. Consequently, many low-income households purchase their groceries from local bodegas and convenience stores, whose products are typically more expensive and less nutritious (Bell et al. 2013).

These barriers to acquiring affordable healthy food are significant because research shows a relationship between healthy food access and health indicators. For instance, populations in greater proximity to healthy food retail establishments have a lower average body mass index as well as lower rates of obesity, diabetes, and diet-related deaths (Treuhaft and Karpyn 2010). Furthermore, not only is there evidence that low-income neighborhoods of color have less access to full-service grocery stores than white communities of any income level, there are also disparities in the "quality, variety, quantity, and price of healthy foods" (Bell et al. 2013, 10). Healthy food access initiatives have creatively addressed this so-called grocery gap through a variety of strategies, including home and community gardens, farmers' market incentive programs, partnerships with food banks, and both mobile and permanent retail approaches.

Grocery Store Attraction

Over the past half-century supermarkets have left city centers for suburban areas, where higher-income consumers, larger land parcels, and more convenient highway access make grocery retail development and operation easier and more profitable. Grocery retailers are particularly sensitive to these considerations because they operate on narrow profit margins of, on the average, less than 2 percent after taxes (Food Marketing Institute 2008; PolicyLink and Local Initiatives Support Corporation 2008). In light of this difficult business climate, retailers or business lenders are often reluctant to enter uncertain markets or to stray from vetted business models. Local, state, and federal incentives now exist to encourage full-service supermarkets to overcome real and perceived challenges—such as crime, concerns about local purchasing power, land acquisition, and financing—by locating in areas with inadequate healthy food retail outlets. Typical grocery store attraction strategies include assistance with assembling financing packages, development incentives, and operating incentives.

Grocery store attraction has become a priority for many areas with limited food access because full-service grocery retailers generally offer a wider range of products at lower prices than smaller grocery stores do (Treuhaft and Karpyn 2010). Some larger grocery stores also have pharmacies, which are badly needed in many underserved neighborhoods. However, Treuhaft and Karpyn suggest that the emphasis on supermarket-based strategies is probably also related to the fact that much of the research on food access has used access to supermarkets as a proxy for access to healthy food in general.

While grocery store attraction incentives can be an important tool for communities with food access problems, overemphasis on this strategy can distract from other approaches that build on existing community assets and engage partners who are more invested in the community's well-being. Anecdotal evidence suggests that in some instances supermarket chains have exploited grocery store attraction programs by locating stores on the edge of a designated incentive zone to take advantage of public incentives while orienting their inventory and marketing to higher-income consumers in adjacent neighborhoods that are already adequately served. The remainder of this chapter explores other healthy food distribution strategies, which include mobile markets, grocery bag subscription programs, buying clubs, retail food co-ops, and healthy corner-store initiatives. One such initiative is the healthy neighborhood market program of the Los Angeles Food Policy Council (see section).

Los Angeles Food Policy Council's Healthy Neighborhood Market Network

OONA MACKESEY-GREEN

The purpose of the Community Market Conversion Program of the Los Angeles Community Redevelopment Agency (CRA) was to improve the availability of healthy foods in select markets and to provide demonstrations of healthy retail makeovers for other neighborhood market operators. The CRA selected four neighborhood markets to participate in its inaugural program in 2012. Eligible stores were required to be owner-operated and located in South Los Angeles. Each participating market was offered a $75,000 forgivable loan for a makeover designed to boost its business and increase its capacity to offer healthy foods in areas with limited grocery access. Shortly after the program launched, however, the California Supreme Court upheld a state law abolishing municipal redevelopment agencies, and the sixty-seven-year-old CRA found itself closing its doors and ending the new Community Market Conversion Program.

Despite the faltering funding, there was already evidence that the program was having an impact. Karen Whitman, the owner of Mama's Chicken on Slauson Avenue, was inspired by the CRA's design work for her store and invested in some modest changes to improve the visibility and availability of the healthy food products she sold. Clare Fox, then the codirector of the Community Market Conversion Program, wanted to honor the CRA's commitment to pilot program participants like Whitman. She partnered with staffers at the Los Angeles Food Policy Council (LAFPC) and experts on its Healthy Food Retail/Food Equity Working Group to host a July 2012 training event for corner-store owners. The daylong event attracted more than one hundred participants and provided training sessions in English, Spanish, and Korean on topics such as financing options for healthy retail food, sources for fresh produce, marketing and management, store design, and CalFresh (food stamps).

The success of the event prompted the LAFPC and the mayor's office to reenvision the Community Market Conversion Program. Positive feedback from training participants suggested that it filled an important niche, but the CRA's demise meant that there were few resources to sustain the program. To Fox, who later became the executive director of the LAFPC, the question became "What can the Food Policy Council do to keep this program alive?"

Operating on a shoestring budget, LAFPC staffers continued to offer training for store operators. They also consulted with individual markets and community organizations interested in launching healthy food retail initiatives. Eventually, the Community Market Conversion Program evolved into the Healthy Neighborhood Market Network (HNMN), an LAFPC program that provided quarterly training and consultation for small-market owners and community organizations managing their own healthy food retail projects. Unlike the Community Market Conversion Program, the HNMN emphasized technical assistance and capacity building rather than providing infrastructure.

The network was built on input from community organizations and partnerships with store owners. Previous program participants became important HNMN champions and valuable resources for new network members. HNMN program associate Esther Park explained, "There is an element of peer-to-peer leadership development and training between the store owners" and noted that store owners who received consultation from the HNMN now share their stories at training events.

HNMN participants identified distribution channels as a challenge for markets that were working to expand their produce selection and locally sourced offerings. The network then connected several store partners with area growers and distributors, including Community Services Unlimited's Village Market Place Program (see separate section). HNMN has also worked with the Leadership for Urban

Renewal Network, a community-based urban redevelopment organization, to create a purchasing cooperative for corner-store operators. Fox explained, "The idea is to scale and aggregate their purchasing power so they are able to interface with larger distributors [and vendors] and get better deals on fresh food products."

Despite positive feedback from store operators, it has been difficult to measure HNMN's impact. Participating markets use different accounting systems, and many have limited sales-tracking technology. However, as HNMN Project Director Daniel Rizik-Baer has noted, defining program success solely in terms of increased sales diminishes the other goals of the initiative. He wants to explore more holistic ways to capture the program's effect by assessing its contribution to changes in the human, intellectual, social, and economic capital of its participants. "The initial outcomes for store owners are knowledge building and skills building," he noted. "It is also community partnerships and relationships." Park added that there's "an understanding [within the network] that this is a transformative process and [requires] a shift in mindset for all the partners that are involved."

Since the closing of the CRA, the LAFPC's collaborative and innovative approach to promoting healthy food retail outlets has attracted new funding from the city's Economic and Workforce Development Department and the Los Angeles County Department of Public Health. This has enabled the HNMN staff to continue its quarterly training and provide direct technical assistance to ten more stores to improve their inventory and merchandising. Furthermore, USDA funding will allow the LAFPC and the the Leadership for Urban Renewal Network to develop a healthy food purchasing and distribution service for corner stores and other local food retailers.

Meanwhile, after her initial investment in enhanced produce merchandising in 2012, Whitman, of the renamed Mama's Chicken and Good Food Market, attended several of the network's training sessions and continued to make improvements to her store. She installed a large open cooler near the store's entrance that highlights a mix of Southern food staples as well as fresh fruit (including citrus), peppers, and avocados, all requested by her customers. In the summer of 2014 she and her family celebrated the store's fiftieth anniversary by offering fresh healthy food to the neighborhood that has consistently sustained her business.

Mobile Markets and Grocery Delivery: Bringing Good Food to Neighborhoods

Mobile vending is another approach to improving healthy food access, especially in neighborhoods that have difficulty attracting supermarkets or have only a few existing grocery retailers (such as ethnic groceries or corner stores). Mobile vendors can offer greater flexibility than brick-and-mortar grocery stores because route frequency, design, and inventory can more easily be modified to align with demand. Although public health practitioners have observed that many mobile food vendors sell unhealthy prepared foods, a growing number of models expressly promote produce and other nutritious food options.

For example, the city of Kansas City, Missouri, promotes healthy mobile vending by offering vendors a 50 percent discount on their annual permit if half the food they sell fulfills specific nutritional standards. If 75 percent or more of a vendor's product offerings meet these requirements, the vendor is given a preferred location. In Oakland, California, and New York City, city health departments worked with other city agencies and mobile vendors to overcome regulatory barriers and expedite permit acquisition to promote produce vending (Change Lab Solutions 2009).

In other cities, nonprofit groups, community development corporations, businesses, and city agencies have partnered in various ways to operate mobile groceries. Although community development corporations are best known for their role in the development of affordable housing, the Central Detroit Christian's Peaches & Greens Mobile Market is an example of how they can also promote healthy food. In Chicago, Growing Power partnered with Mayor Rahm Emanuel to operate a refurbished Chicago Transit Authority bus as the Fresh Moves Mobile Market in July 2015, with plans for additional buses. These mobile markets sell Growing Power produce and a limited selection of wholesale-sourced produce. Fresh Moves makes three or four stops per day at community-based partner organizations; eventually, up to twenty-two distinct market stops will be made weekly, bringing consistent access to high-quality produce to two thousand families per month. Fresh Moves also offers nutritional information and recipes to build farmer-consumer relations and to promote fresh food awareness and knowledge among its customers.

Finally, Internet grocery delivery services such as PeaPod represent another grocery distribution strategy potentially suited to households with limited transportation access. However, since these services require Internet access to place orders and may not accept federal food stamps, barriers to low-income households are still present.

Improvement of Food and Transit Linkage

Recognizing that the lack of transportation can be a barrier to healthy food access, cities are increasingly exploring ways to better align public transit routes and schedules with grocery stores and farmers' markets. For example, in Austin, Texas, the local food policy council and nonprofit partners conducted interviews with area residents to assess food access needs and identify desired grocery routes. The groups subsequently worked with the Capital Metropolitan Transit Authority and area supermarkets to develop a grocery bus that links an underserved Latino community with two full-service supermarkets. In Knoxville, Tennessee, Knoxville Area Transit worked with area supermarkets to develop the Shop & Ride program, in which customers are eligible for free return tickets in exchange for purchasing at least ten dollars of groceries. Participating supermarkets reimburse the transit agency for fares associated with the program on a monthly basis (Ringstrom and Born 2011). Akron, Ohio, and Hartford, Connecticut, have also developed food, retail, and job-oriented public transit routes (Metro Regional Transit Authority n.d.; Vallianatos, Shaffer, and Gottlieb 2002).

In areas with poor public transit systems or where ridership is not sufficient to justify new transit routes, privately and publicly sponsored shuttle programs may offer an alternative. A California feasibility study on privately operated supermarket shuttle programs found that "shuttle programs improve customer loyalty, reduce costs from shopping cart loss and retrieval, and win new customers" and that such shuttle services could operate at or below cost with a per-trip minimum of $25 per customer (Mohan and Cassady 2002, 4). Nevertheless, many large retailers remain wary of the costs and logistics associated with operating shuttle programs and have deferred to nonprofit groups and city agencies to fund and operate such services. For example, in Madison, Wisconsin, a subsidized grocery taxi service program was implemented after the closure of the last grocery store in an area with low food access. However, because of its cost, this program was intended to serve only as a stopgap measure. At the time of this writing, a community group was working with the city and a local natural foods cooperative to develop a retail cooperative in the neighborhood.

TOWARD INTEGRATIVE AND SYSTEMIC SOLUTIONS

Community and regional food distribution should be built into markets that reward local and regional producers and their supply chain partners for practicing sustainable agriculture, equitable supply chain governance, and resource-efficient transportation and logistics. Conversely, they should provide equitable access to healthy food. The tension between these two objectives is epitomized by the *fair pricing dilemma*, the notion that distribution models "designed to help producers retain a larger percent of the retail dollar typically operate at price points that make their products unaffordable to low-income markets" (Day-Farnsworth, Zimmerman, and Daniel 2012, 3). The final two sections in this chapter offer examples of marketing and retail initiatives that strive to simultaneously address these objectives by building markets for sustainably grown products and increasing healthy food access in communities with limited healthy food retail outlets.

In these sections, Community Services Unlimited and the Detroit Black Community Food Security Network describe the slow and thoughtful evolution of two urban agriculture and marketing operations that are as intentional about sourcing from small family farms and from farmers of color as they are about ensuring that their operations serve communities with limited healthy food retail outlets. Both organizations view their food production and distribution projects as part of their larger mission of community empowerment and development. For these organizations the strategy is as important as the outcome. The objective is not simply to increase access to healthy food but to increase access in a way that, as Community Services Unlimited's Dyane Pascall says, "uplifts" the community. As these examples demonstrate, this means improving food access while also creating jobs and community spaces that foster activities such as recipe sharing, cooking skills, information exchange, and relationship building. These holistic and community-driven approaches to food supply chain development are encouraging, but the economic viability of affordable, good-food marketing and distribution operations remains uncertain. Both organizations in the examples have benefited from close affiliation with nonprofit groups that have attracted grant funding; the question is whether such businesses can succeed without grant subsidies or significant improvements in economic conditions within the communities they serve.

Community Services Unlimited's Village Market Place

OONA MACKESEY-GREEN

Founded in 1977 by the Southern California chapter of the Black Panther Party, Community Services Unlimited (2016) has a long history of "supporting and creating justice-driven community-based programs and educational initiatives." Because issues such as food production and food access are at once tangible and systemic, CSU found that food-related projects were a natural fit for an organization that was working both to address immediate community concerns and to cultivate critical consciousness and long-term community transformation. By 2005 CSU had established two youth agricultural programs and minifarm sites throughout South Los Angeles.

In 2007 the organization expanded to include produce marketing in order to increase its financial self-reliance while improving healthy food access in South Los Angeles. CSU's Financial and Administrative Manager Dyane Pascall recalled, "It came out of the need of working with folks in the community and knowing we couldn't just be sitting here and saying, 'Eat healthy,' but we have to actually make produce available while creating jobs and revenue for our programs." Initially CSU operated a produce stand and produce bag subscription program at its Expo Urban Mini Farm, reaching about eighty families and earning $7,000 in sales during its first year. In time this became one of CSU's cornerstone programs: the Village Market Place.

CSU initially sourced produce for the program from its two urban minifarms and supplemented it with produce from local farmers' markets. The demand increased as local organizations requested CSU-sourced produce at their own community outreach sites. One of CSU's early partnerships was with the health-care community-organizing group Community and County Health Empowerment (C+CHE), which approached CSU to supply a produce stand at the Los Angeles County–USC Medical Center. "It was early on in their organizing," explained CSU Associate Director Heather Fenney Alexander, so "they were looking to be strategically present in the community and have a venue for talking to people about health-care issues." CSU eventually became the produce stand manager, and C+CHE continued to do community outreach through the stand.

FIGURE 7. The Village Market Place produce stand at Mercado La Paloma in South Los Angeles. (Courtesy of Anne Pfeiffer, CRFS)

CSU has started supplying produce to several community markets in South Los Angeles, including Mama's Chicken and Good Food Market (mentioned in an earlier section) and the community market space Mercado La Paloma (fig. 7). Although these partnerships are not very profitable for CSU, the group recognizes that it fills an important gap by distributing small volumes of products, which few large distributors are willing to do. "With the corner stores in particular," Alexander noted, "we are offering a pricing and a volume that is manageable for them to take. We do something unique with the corner stores that we wouldn't do for everyone" because they are an important source of fresh produce for the neighborhood.

Although much of its work focuses on communities in South Los Angeles, CSU's relationships with regional farmers have also grown over the years. CSU has maintained many partnerships with its original farmers' market suppliers and works with new growers as well. Its regional producers are not required to be certified organic or to meet special labor or production standards; quality assurance is built largely on close communication and trust. According to Alexander, "We start small and make small purchases. We have a questionnaire that we do with the farmers to get a sense of [their farms and practices]."

In 2015 CSU acquired the Paul Robeson Community Center on South Vermont Avenue as its future headquarters. Although the long-term vision for the center will take time and funding to realize, experience suggests that CSU's approach to relationship building and community-driven programming are crucial to provide integrated strategies that improve healthy food access for South Angelenos, build markets for regional growers, and facilitate critical consciousness in the community.

With community-based economic development, Pascall explained, you have to ask yourself, "Is [this] leading to the uplifting of your community or the demise of your community? [The Village Market Place] came from a lack of . . . businesses that uplift the community, that have a long-term commitment to the community by way of food. It's one thing to be a food business, but it's another thing to be responding to the needs of the people. That is what our existence is about; food is definitely the tool that we use."

The Detroit People's Food Cooperative

NICODEMUS FORD AND MALIK YAKINI

Food cooperatives possess historical and cultural significance to underserved communities. A corporation that extracts wealth from the community may own traditional and larger grocery stores. Cooperatives, in contrast, are grassroots examples of reorganizing capitalist structures by enacting democracy for the collective benefit of the local residents rather than individual profit-driven monopolies. In an era of concentrated wealth, food cooperatives possess the power to dismantle an inequitable system that prioritizes profits over people.

The Detroit People's Food Cooperative (DPFC) was in development as of the summer of 2016. It will be a community-owned grocery store in Detroit's North End neighborhood. The DPFC grew out of the Detroit Black Community Food Security Network's Ujamaa Co-op Buying Club, where for several years members could purchase a wide range of healthy foods, supplements, and household items at discounted prices. Detroiters' need for greater access to food grown and distributed by and for the community drove efforts to expand the buying club into a brick-and-mortar retail store. Several community outreach activities were conducted to learn about customers' needs, buying habits, and location. Consumers' needs were identified early in the process of designing a store by and for the people of Detroit. The North End is in close proximity to the city's main thoroughfare, Woodward Avenue, and provides fairly easy access to local highways. This location ensures that the city's most disadvantaged citizens can have greater access to local healthy food.

The goal is to source from as many local growers as possible—particularly black producers—to circulate dollars in the community. The Detroit Black Community Food Security Network and its D-Town Farm will contribute produce sold at the cooperative. We seek to partner with other producers who share in our cause to transform and empower Detroiters.

DPFC customers and co-op members will benefit from purchasing local fresh food. The members will own their cooperative. In a city that has largely seen wealthy capitalist interests trump the common will of the people, the co-op embodies self-realization and empowerment for Detroiters. It will serve as a community hub, providing training, education, and information to the members and residents, particularly youth.

Residents, many of whom are considered low-income, can be partial owners of the store. Co-op members are owners of the store and the main beneficiaries. In addition, the store will employ about twenty local residents. The DPFC also seeks to expand beyond the traditional functions of a grocery store. Black people need not only access to healthy and fresh food but also the knowledge to cook, grow, and build community with one another. In this way the DPFC will serve as a community hub, with cooking and nutrition classes and demonstrations, public board meetings, and informational sessions. The elements of a triple bottom line—economic return to the member owners, sustainability, and care for the community—are inextricably linked.

CONCLUSION

Community and regional food distribution and supply chain development highlight the diverse challenges that surface when we intentionally try to build values like justice and fairness, strong communities, and vibrant local economies into the food system. On the production side, the challenges relate primarily to market access and the specific mechanism through which values are embedded in community and regional supply chains. With technical assistance and through experimentation, small to midsize sustainable farms and their supply chain partners are overcoming market access challenges by expanding their production capacity, increasing their knowledge of best practices in marketing and supply chain governance, and improving the efficiency of their delivery systems.

On the consumer side, distribution challenges relate to healthy food access. The goal here is to increase the availability, affordability, and accessibility of fresh healthy food options in communities without adequate food retail outlets. Finally, several of the case studies in this chapter highlight how food distribution has the potential to be about much more than moving products to consumers. By considering food distribution operations within the context of community transformation, we see examples of how they can also be mechanisms for creating jobs, building relationships, and strengthening and empowering communities.

CHAPTER FIVE

Food Processing as a Pathway to Community Food Security

GREG LAWLESS

In this chapter Greg Lawless makes important points about the potential of food process-
ing in promoting an adequate diet of fruits and vegetables among low-income residents
of American cities through examples of initiatives in Los Angeles and Wisconsin. He
suggests roles for local processors to address the supply-and-demand constraints that
limit the consumption of fruits and vegetables—the food processing component of the
food system supply chain.

E very year the federal government reports that most Americans of *all* economic strata do not consume the recommended daily quantities of fruits and
vegetables. To address that deficiency in low-income urban neighborhoods,
policy makers, community organizations, farmers, and others have supported
strategies to make locally grown fresh fruits and vegetables more accessible and
affordable.

This chapter identifies the constraints on both the supply of local produce
and the demand for fresh produce that limit the effectiveness of these well-
intended efforts. It considers how food processing—the safe transformation of
plant and animal materials into products for sale to consumers—can overcome
these supply-and-demand constraints. Finally, by comparing four food processing strategies in Los Angeles and Wisconsin, the chapter offers preliminary
conclusions about the potential of food processing as a pathway to community
food security.

INADEQUATE FRUIT AND VEGETABLE CONSUMPTION

The dietary guidelines published in 2015 by the USDA and the US Department of Health and Human Services (HHS) (USDHHS and USDA 2015) recommended one to two and a half cups of fruit and one to four cups of vegetables per day depending on one's age, sex, and level of physical activity. The guidelines further stated that these recommended daily servings can be met by eating fresh, frozen, or canned produce or meals and snacks with fruit and vegetable ingredients—provided that added fat, sugar, and salt do not exceed the daily maximum limit.

The Centers for Disease Control and Prevention within the HHS promotes the consumption of fruits and vegetables to ensure an adequate intake of essential nutrients, to reduce the risk of cardiovascular disease and certain types of cancer, and to maintain a healthy weight. However, only a small percentage of Americans meet the recommendations for eating fruits (13.1 percent) and vegetables (8.9 percent) (Moore and Thompson 2013).

Using somewhat different intake measures, the Center for Urban Population Health (a partnership of the University of Wisconsin and Aurora Health Care) showed almost no difference in fruit and vegetable consumption between the lower and middle socioeconomic groups in Milwaukee (table 3; Greer et al. 2014). Even though neighborhoods representing the highest levels of income and education fared somewhat better, only 37 percent of the city's wealthier residents met the federal guidelines.

INCREASING THE ACCESSIBILITY AND AFFORDABILITY OF LOCALLY GROWN FRESH PRODUCE

All seven research cities in the CRFS project have ongoing efforts by activists and entrepreneurs to improve food security, health, and nutrition in urban communities by making locally grown fresh fruits and vegetables more accessible and affordable. For example, community gardens and commercial urban farms in Boston and Los Angeles are increasing the local supply of fresh produce in low-income communities. Health and consumer advocates in Milwaukee, Chicago, and Detroit use federal nutrition assistance programs to help customers purchase these products through farmers' markets and other marketing channels.

Hmong and Latino farmers of the Spring Rose Growers Cooperative outside Madison, Wisconsin, combine production and marketing approaches as they pro-

TABLE 3. Percentage of Population Consuming Five or More Servings of Fruits and/
or Vegetables per Day

Consumer Population	% Eating Adequate Fruits and Vegetables
Lower socioeconomic status in Milwaukee	30.3
Middle socioeconomic status in Milwaukee	30.8
Higher socioeconomic status in Milwaukee	37.1
Total population in Milwaukee	31.5

mote their CSA produce at subsidized prices in locations where their low-income neighbors pick up monthly Women, Infants, and Children (WIC) program coupons. In a similar fashion, Matthew 25, a faith-based nonprofit organization in Cedar Rapids, Iowa, grows vegetables on former residential lots ravaged by the flood of 2008 and distributes the harvest through CSA shares to its low-income neighbors at half the standard market rate.

Such efforts have a real and positive impact on people's lives and certainly merit continued support. However, there are inherent constraints to using only locally grown fresh fruits and vegetables as a means of meeting daily nutritional recommendations and reducing food insecurity. Specifically, we can identify two barriers related to local supply and four barriers related to consumer demand. Once we understand these constraints, we can consider whether and how food processing strategies can help to overcome them.

CONSTRAINTS ON THE LOCAL SUPPLY OF FRESH FRUITS AND VEGETABLES

One supply challenge associated with locally grown produce is *seasonality*. Aside from Los Angeles, farmers in our other project cities contend with long cold winters. Although innovative growers respond to this challenge with season-extending techniques such as hoop houses, floating row covers, and indoor growing systems, thus far they have been unable to achieve the variety, scale, and lower production costs attainable in warmer months.

Of course, the problem of seasonality is overcome in northern cities by importing produce throughout the winter. However, advocates of local food resist this solution for legitimate reasons, including environmental sustainability, support for family farms, and the retention of dollars in the regional economy. One effect of seasonality is that the prices of imported fresh produce in northern climates can be considerably higher than those during the abundance of local harvests, and this price difference affects low-income consumers more than other groups.

Another constraint related to locally grown fruits and vegetables is the *limited acreage in production* in and around most metropolitan areas, which can result in an insufficient supply to meet the demand. Within this discussion we must also note a distinction between produce grown for fresh markets and produce grown under contract for industrial processors.

Some varieties of fruits and vegetables are produced under conditions that make them more suitable for retail sale as fresh products to households, restaurants, and cafeterias. Taste and appearance are more desirable qualities in these markets. Other varieties and growing methods are better suited for large shipments to processing facilities for canning and freezing or for use as ingredients in manufactured food products. In selecting seed varieties for these markets, taste and appearance may be sacrificed in favor of higher yields and lower costs of production, handling, manufacturing, and distribution.

Wisconsin provides an example of these distinctions. Using the Impact Analysis for Planning (IMPLAN) input-output model, a recent University of Wisconsin Extension report estimated a $100.5 million market value of consumer demand for fresh produce from households and institutions in the seven-county region around Milwaukee in 2012 (Kures 2013). In contrast, the same report considered the supply of fresh produce, using USDA production statistics for twenty-four selected fresh vegetables and melons, and found that Wisconsin farmers were responsible for just 0.3 percent of the country's production of these crops for fresh markets in 2012. Cabbage, onions, and sweet corn topped the list, with a total harvest value of all fresh market produce at $32.9 million.

The UWEX report also used IMPLAN to estimate $47.9 million of industrial demand in the Milwaukee region for eight vegetable crops. In terms of supply, Wisconsin ranked second in the nation in the production of these crops for fresh markets in 2012, with a harvest that the USDA valued at $196 million.

We should note that agricultural supply (using the prices paid to farmers) cannot be accurately compared to the demand by households and industry (using retail and wholesale prices, respectively) without factoring in the costs for transportation, storage, marketing, and profit among the various intermediaries. Nevertheless, it seems apparent that the demand for fresh produce from the seven counties around Milwaukee far exceeds the supply from the rest of the state, whereas the statewide supply of processing vegetables far exceeds industrial demand in the seven-county region.

The mismatch of supply and demand for fresh produce is not news to nutrition and local food activists in Milwaukee, who struggle to attract vendors of fresh produce from local farms into low-income urban communities. Basic economic theory dictates that when a product is in short supply, it is allocated to higher-paying customers, at least in the absence of subsidies or programs that alter conventional market dynamics.

Public policy advocates may offer a wide range of ideas to increase the production of fresh produce and its availability and affordability in low-income communities. At the federal level, the USDA has programs targeted to local and regional food systems, farmers' markets, farm-to-school initiatives, sustainable and organic research, beginning and socially disadvantaged farmers, and many others (Fitzgerald, Evans, and Daniel 2010). The National Sustainable Agriculture Coalition is an excellent source of information about these programs.

Many states have similar programs, such as the support of farmers' markets, farm-to-school programs, and various subsidies to increase the purchase of healthy foods through nutrition programs such as WIC and SNAP. However, policy change takes time; it requires farmers and distributors who are willing to work through the challenges of grant applications once the programs are in place. (For more on policy innovations, see chapter 12.)

Fortunately, just as fruit and vegetable farmers are overcoming the constraints of seasonality through innovative production practices, urban-based farms are overcoming limited acreage in production through innovative arrangements with corporations and churches and by using foreclosed land and public open spaces to make more land available, often for free or at little cost. This expanded acreage and subsidization can have a direct effect on the availability and affordability of fresh fruits and vegetables in low-income communities.

CONSTRAINTS ON THE DEMAND FOR FRESH FRUITS AND VEGETABLES

For decades researchers on nutrition and behavior have explored why Americans do not eat enough fruits and vegetables, with many studies focusing on low-income communities of color. Following are some examples of studies in the past twenty years:

- Low-income women in Maryland reported that major barriers included "not liking specific fruits and vegetables, preferring other foods, time and difficulty involved in preparation, perishability, and cost" (Treiman et al. 1996).
- African American men and women in north-central Florida reported that their busy lives did not give them much time to prepare healthy meals, and there was a common perception that healthy foods do not taste good (James 2004).
- Low-income African American mothers in St. Paul–Minneapolis identified four barriers to serving and eating more fruits and vegetables, in the following order of importance: taste or liking, cost, preparation time, and spoilage (Henry et al. 2006).
- African American men and women in Philadelphia ranked preferences and likes, cost or finances, availability and convenience, and taste or flavor as the most salient barriers to fruit and vegetable consumption (Lucan, Barg, and Long 2010).
- Low-income adults in North Carolina listed cost, transportation, quality, variety, changing food environments, and changing social norms as barriers. The last barrier particularly affected women in the workforce, who cited "lack of time during the day to cook, the convenience of prepared or fast foods, and the struggle to provide food that their children liked" (Haynes-Maslow 2013).

It was noted earlier that many community organizations and activists work to make locally grown fresh fruits and vegetables more accessible and affordable for low-income urban consumers. However, these studies indicate that some barriers to consumption may actually be a result of the nature of fresh produce itself. In other words, fruits and vegetables appear not to be a greater part of people's diets in some cases because too many consumers (1) dislike their taste, (2) have limited time to incorporate them into flavorful meals, (3) want to avoid the cost of spoilage, and (4) see them as too expensive.

These four common complaints—taste, time, spoilage, and cost—present a challenge to nutrition advocates who promote fruit and vegetable consumption as part of a healthy diet. Sustained efforts to educate consumers about growing, storing, and cooking these products can have a positive effect. To complement such efforts while recognizing the persistent failure of so many Americans to eat the recommended daily portions of fruits and vegetables, food processing offers another potential pathway.

FOOD PROCESSING STRATEGIES IN LOS ANGELES AND WISCONSIN

Commercial food processing is defined as the variety of practices used to transform plant and animal materials (such as grains, produce, meat, and dairy) into products for sale to consumers in compliance with associated food safety regulations (Johns Hopkins Center for a Livable Future 2010).

A stigma is often associated with the term *processed food*. As food system critic Michael Pollan (2008) has advised, consumers should shop at the periphery of a supermarket and avoid the middle aisles, where processed foods, labels with dubious health claims, and products with unpronounceable ingredient names predominate. Although many nutritionists might agree with that sentiment, our pursuit of community food security requires a broader view that accommodates demand constraints. Commercial food processing can occur at every level: private homes, start-up companies, nonprofit initiatives using shared kitchen facilities, midsize family-owned businesses, and major industrial manufacturers. To overcome the challenges associated with locally grown fresh fruits and vegetables, innovation should be supported at all these levels.

As the CRFS project began exploring food systems in the project cities, we looked for innovations in food processing that improved nutritional outcomes and reduced food insecurity in low-income communities. Some organizations came close to this dual goal, including CropCircle Kitchen in Boston (recently renamed CommonWealth Kitchen) and Food Enterprise & Economic Development (FEED) Kitchens in Madison, but they were primarily oriented to entrepreneurship and product development, not food insecurity. Eventually we identified four initiatives—two in Los Angeles and two in Wisconsin—that would serve as case studies to evaluate the potential for food processing to overcome the barriers to supply and demand for fruit and vegetable consumption.

The Farmer's Kitchen, Los Angeles

The CRFS project approached Sustainable Economic Enterprises of Los Angeles (SEE-LA) because of its history of introducing farmers' markets into low-income

FIGURE 8. The Farmer's Kitchen in Hollywood, with former executive chef Ernest Miller. (Courtesy of Greg Lawless, CRFS)

neighborhoods. It had recently built a food processing facility with the stated purpose of linking the urban population with small California farms (Organic Authority 2012). After seven years of planning, SEE-LA opened the Farmer's Kitchen in 2009 to complement its flourishing Hollywood Farmers' Market. The ambitious mission statement included two goals that are pertinent to supply and demand for fresh produce. The kitchen would "promote sustainability of California small farmers" by incorporating foods from the adjacent market and "improve access to fresh produce to improve nutrition and reduce hunger" (Farmer's Kitchen 2014).

On visiting the Farmer's Kitchen in 2012, the CRFS project team learned that cooking and food preservation were major activities. Pickled and preserved produce grown on local farms was displayed on retail shelves in glass jars, and the executive chef offered training in the Master Food Preserver Program through the University of California Cooperative Extension (fig. 8).

In terms of demand constraints, preserving fresh food in the household can improve the taste of fruits and vegetables and reduce spoilage. However, it also requires significant time—something that studies indicate low-income people lack. Furthermore, the potential for the Farmer's Kitchen food preservation classes to help low-income residents was probably limited, because the immediate area had gentrified considerably in the twenty-plus years since the Hollywood Farmers' Market had opened (Karp 2012).

In terms of supply constraints, seasonality is not a major issue in Southern California. Nevertheless, the potential for the Farmer's Kitchen to handle large volumes and increase the demand for fresh produce from local farms appeared limited by its scale of processing products like red-cabbage sauerkraut. That is, the small size of its facility and its rudimentary equipment had little influence on the crop and marketing choices of local growers.

Although the Farmer's Kitchen was innovative in its integration with a thriving farmers' market, its ability to overcome the demand barriers associated with locally grown fresh fruits and vegetables appeared to be quite limited. In the summer of 2014 the Farmer's Kitchen ceased operations, and the facility was transformed into a privately owned restaurant called Field Trip, which offered local seasonal produce and catered to an upscale lunch crowd until it too closed, in 2015.

Homeboy Industries, Los Angeles

On the same trip to Los Angeles in 2012 we visited Homeboy Industries, a nonprofit organization that addresses the needs of former gang members and previously incarcerated men and women. Founded by a Catholic priest in 1988 through a project called Jobs for a Future, the organization attained nonprofit status in 2001.

Homeboy Industries has its headquarters in a commercial area of Chinatown and provides job training, income, and a sense of purpose for employees through several social enterprises. Its food-related ventures include Homegirl Café, which offers full-service breakfast and lunch; a bakery; a catering service; a small diner in Los Angeles City Hall; and a take-out location at Los Angeles International Airport.

Although we saw a retail display of Homegirl juices and seasonal preserves at the headquarters building (fig. 9), the bulk of commercial activity from Homeboy's food enterprises comes from the restaurant and catering operations rather than food processing. However, this approach can help de-emphasize the boundary between these traditionally separate segments of the food industry, especially for locally grown fresh produce.

Restaurants and food processing outlets can transform raw fruits and vegetables into more appealing foods and reduce cooking and other food preparation time for

FIGURE 9. Homeboy Industries of Los Angeles and its branded Homegirl juices. (Courtesy of Greg Lawless, CRFS)

busy consumers. When restaurants and food processing outlets purchase these ingredients from local farmers, they can incentivize the expansion of agricultural production. Small-scale food processing outlets and restaurants with an orientation to local foods can use some of the same labor-saving equipment. Although the two generally operate under separate sets of food regulations, there is overlap in some of the measures that promote food safety, like food handling and storage. Finally, both restaurants and food processing outlets have the potential to provide healthier food options to low-income consumers.

However, the food enterprises of Homeboy Industries developed to create jobs and employment training for the high-risk populations it serves. As such, it is not primarily oriented either to local farmers or to low-income consumers. The organization does engage in limited urban farming. Its website states that the goal of the minifarms pilot project is to provide as much as 30 percent of the produce and herbs it needs to supply its restaurant and bakery (Homeboy Industries 2013). However, at the time of our visit the restaurant menu provided no indication that Homeboy Industries sourced the rest of its produce directly from local farms.

In terms of the customer base, the organization's cafés and bakeries are in commercial areas that cater more to middle- and upper-income consumers. None of the farmers' markets where Homeboy Industries vends its products appear to be in the "concentrated poverty neighborhoods" clustered throughout South Los Angeles and adjacent to downtown (Matsunaga 2008).

In 2013 more than 250 former gang members and people with past gang affiliations were employed full-time in the seven enterprises of Homeboy Industries, through job training programs that prepared them for life and employment outside gangs and prisons (Homeboy Industries 2013). Although the wages improve the employees' ability to afford healthy food, addressing food insecurity is not the organization's primary mission.

The two organizations we visited in Los Angeles were inspiring and impressive, but their food processing activities were quite limited, the volume of local produce used was fairly minimal, and the consumers served generally did not live in low-income communities. Neither the Farmer's Kitchen nor Homeboy Industries appeared to address the demand barriers associated with local fresh fruits and vegetables.

Field to Foodbank, Madison

The Field to Foodbank initiative began in 2010 when a large-volume carrot grower with an overabundant harvest was forced to let forty acres of carrots perish in the field because his local processing plant was over its capacity. That experience led to a discussion with UW–Madison horticulture professor Jed Colquhoun (fig. 10) and other growers about diverting excess crops to the Second Harvest Foodbank of Southern Wisconsin the following year (Fischer 2012). As a member of the national organization Feeding America, the Madison-based Second Harvest handles about thirty-one thousand pounds of food each day and distributes more than twelve million meals annually through its partner agencies and programs.

The innovation of Field to Foodbank is to work through the logistics of communication and transportation to connect growers, processors, distributors, and the food bank in getting produce to clients. To avoid the waste of perishable donations and the costs of refrigeration and freezing, farmers' excess produce is canned and stored by an industrial canner, then shipped over time by distribution partners to the Second Harvest facility as needed. Donations to the program primarily consist of carrots, potatoes, onions, sweet corn, snap beans, and apples. The produce, cans, processing service, and trucking are donated in whole or in part.

FIGURE 10. Jim Scheuerman (left), formerly of Second Harvest, and Jed Colquhoun of the University of Wisconsin–Madison. (Courtesy of Kris Tazelaar, Second Harvest Foodbank)

The Field to Foodbank initiative partly addresses the supply constraint of seasonality in a northern climate by finding an outlet for excess produce that might otherwise be lost to frost. However, an opportunity for charitable giving is not a market incentive for growers to expand production, nor does it mitigate the financial risks associated with overproduction. In fact, many of the growers involved have not enjoyed a tax benefit because they took their allowable deductions in other ways. Nevertheless, the participating farmers derived enough personal satisfaction from the early success of the program to dedicate a small portion of their crop to the food bank system, rather than simply hope for above-average yields.

Even though a few tons of sweet corn is not a large sacrifice from a one-hundred-acre crop, such donations enable the food bank network to provide vegetables to low-income consumers and to allocate its limited resources to purchase other foods. Because canned foods do not spoil, the risk of waste is eliminated, both at the food pantry and in the consumer's home. Already cleaned and chopped, canned vegetables save the consumer a modest amount of preparation time. However, the taste of canned foods may not appeal to some consumers, which may limit the program's effect on fruit and vegetable consumption in food banks.

Growing Power's Carrots-to-Schools Project, Milwaukee

Will Allen's nonprofit organization Growing Power grows and distributes food in Milwaukee and Chicago while training and inspiring hundreds of aspiring urban and rural farmers to expand what Allen (2012) calls the Good Food Revolution. Allen also served as codirector of the CRFS project. During a project gathering in 2011 Allen reached out for help in implementing what he saw as an opportunity: Milwaukee Public Schools, the largest school district in Wisconsin, was interested in purchasing two hundred thousand pounds of carrots—a crop he had limited experience growing. Over the ensuing years, the effort expanded to involve two prominent Milwaukee companies, a large rural carrot producer, a prominent organic grower, several university specialists, an innovative local food broker, and two major public school districts.

At the conclusion of the 2013 growing season, which featured plentiful rain and the acquisition of labor-saving equipment, Allen (fig. 11) proudly shipped thirty-six thousand pounds of fresh coin-cut (i.e., sliced in round pieces) carrots to Chicago in what the USDA called the largest farm-to-school transaction in US history (Healthy Schools Campaign n.d.).

As significant as the accomplishment was, however, the enormous challenge of community food security in major urban centers remained apparent: Growing

FIGURE 11. Growing Power's Will Allen harvests carrots for Chicago public schools. (Courtesy of Greg Lawless, CRFS)

Power's eighteen tons of carrots could supply the district's four hundred thousand children for only one day (Shannon 2014). Nevertheless, Allen's success in shipping the carrots to Chicago Public Schools stemmed in large part from keeping the product's price low enough for the school district to afford. Many factors contributed to the low price, including the cooperation of experienced and efficient businesses that processed, shipped, and distributed the product after harvest. The cost of production was also subsidized by the rent-free contribution of fertile peri-urban farmland owned by Sysco Corporation outside Milwaukee. The value of other subsidies is more difficult to estimate, because as a nonprofit organization Growing Power receives many grants, donations, and volunteer support that sustain its operations.

Growing Power's Carrots-to-Schools Project is similar to the Field to Foodbank initiative in several ways, including cooperation among farmers, corporate generosity, university support, and the use of existing supply chains. However, unlike the Field to Foodbank producers, Allen did not start his experiment with decades of experience in large-scale carrot production or with knowledge of processing and storage options. Part of his nature as an innovator is to charge into unknown

	Impact on Local Supply of Fruits and Vegetables		Impact on Local Demand for Fruits and Vegetables			
	Seasonality Constraint	Acreage Constraint	Taste Constraint	Time Constraint	Spoilage Constraint	Cost Constraint
Farmer's Kitchen	Low	Low	High	High	Low	Low
Homeboy Industries	Low	Low	High	High	Low	Low
Field to Foodbank	Moderate	Low	Low	Moderate	High	High
Growing Power	Low	Moderate	Low	Moderate	Low	High

territory, with consequent periods of trial and error. As a result of the initiative, he and his staff developed new capacities that will help them expand and improve their efforts each year. Another notable difference is that whereas the Field to Foodbank project involves canned fruits and vegetables, the Growing Power carrots were sliced into coins by a processing partner in Milwaukee and served fresh to Chicago students. It is not known if the children preferred the fresh carrots to a cooked alternative; we will return to the essential matter of taste appeal in the analysis and recommendations to follow.

COMPARISON OF EFFECTS ON SUPPLY- AND-DEMAND CONSTRAINTS

This chapter began by presenting two supply constraints (seasonality and limited acreage in production) on locally grown fruits and vegetables in and around cities and four demand constraints (taste, time, spoilage, and cost) on low-income consumers' concerns about fresh produce. Table 4 offers a cursory estimation and comparison of how effectively the Los Angeles and Wisconsin examples overcame these six constraints.

Los Angeles

The original intent of the Farmer's Kitchen was to use fresh produce and other farm products from the adjacent farmers' market. The CRFS project team visited the facility under the impression that shelf-stable processed food products would be a substantial component of the operation. We learned, however, that the scale of food preservation was quite small and that the processed products the Farmer's Kitchen developed were not marketed beyond a small retail area. Instead the enterprise functioned more as a restaurant serving breakfast and lunch. Although seasonality is hardly a constraint in Southern California, most of what the Farmer's Kitchen produced was prepared to be eaten immediately, except for small quantities of fermented and canned products. We did not determine how much local produce the restaurant purchased on an annual basis, but it was probably insufficient to encourage many farmers to expand their acreage in production.

The scale of food processing for Homeboy Industries may be more extensive than that of the Farmer's Kitchen. The company offers juices and preserves and recently introduced a line of chips and salsa. It also has permanent outlets in several locations and participates in farmers' markets throughout the region. We saw no indication that Homeboy Industries factors the source of its produce into menu planning for its cafés and bakery, beyond the use of its own minifarm production. Of course, much of the produce comes from the abundant year-round farms in the area.

Overall, both Los Angeles examples have had little influence on supply constraints. In terms of their effect on demand constraints, both offered flavorful, appealing meals with ample portions of fruits and vegetables. Also, their customers were spared time spent preparing meals by dining out. However, eating out is a luxury for people on a limited income.

In terms of costs, a sandwich, salad, and beverage at the Farmer's Kitchen could be purchased for as little as $7.50 in 2014. In comparison, a "combo meal" at a Los Angeles McDonald's costs an average of $6.99. Since the Farmer's Kitchen ceased operation in 2014, perhaps its practice of using local farm ingredients while charging fast-food prices was not sustainable. The cheapest sandwich, salad, and beverage on the Homegirl Café menu cost $13.50 in 2014. Both restaurants are or were located in Los Angeles neighborhoods in which high percentages of the populations live on median household incomes of less than $20,000 per year (Hollywood, 39.4 percent; Chinatown, 53.6 percent). However, both ventures opened in safe and attractive commercial districts surrounded by new development. The prices necessary to sustain higher-quality restaurants in such locations could be prohibitive for low-income families in the areas (*Los Angeles Times* 2014).

Although the Farmer's Kitchen, which had set out to serve both small rural farmers and a low-income urban population, is no longer operating, the annual impact of Homeboy Industries on thousands of clients can hardly be overstated. The former gang members and inmates the organization hires are developing critical job skills. There is no indication that the organization ever intended for its food enterprises to address farm or food access issues, so we will not appraise their efforts in those terms.

After evaluating the two Los Angeles examples for their effects on demand constraints related to local fresh fruits and vegetables, it appears that neither makes a strong case for how food processing might provide a pathway to community food security. However, it may be worth considering how effectively restaurants can make healthy food flavorful and appealing while saving time for consumers who lack it.

Wisconsin

The two Wisconsin examples are more effective in illustrating the promise and challenges of food processing as a means of addressing demand constraints. By canning vegetables the Field to Foodbank project clearly extends their availability long after harvest. Canned products also avoid the spoilage that can constrain consumer demand. Growing Power's freshly cut carrot coins, in contrast, would have little influence on seasonality or spoilage because they remain perishable products, even though carrots do store longer than many vegetables.

Because the Field to Foodbank project provides no income to the participating farmers, it offers no monetary incentives to significantly expand acreage in fruit and vegetable production. Nevertheless, the small increases that some large farms make in dedicating future crops to the program are significant to the food bank agency they supply. Growing Power, in contrast, took a significant step to increase production when Milwaukee Public Schools expressed interest in purchasing two hundred thousand pounds of carrots just before the 2011 growing season. Despite having no experience with that crop, Growing Power would eventually plant fourteen acres in preparation for meeting the demand from clients in Milwaukee, other Wisconsin schools, and Chicago.

In terms of demand constraints, the CRFS project did not determine whether the food bank clients or students enjoyed the taste of the canned and freshly cut products provided by Field to Foodbank and Growing Power, respectively. However, if we accept the studies that show many consumers' dissatisfaction with the flavor of fruits and vegetables, it is hard to conclude that the minimally processed products in these two initiatives had much effect on taste constraints. Yet a consumer can easily open a can at home and a school cafeteria employee can easily serve washed and freshly cut carrots, so each was saved a moderate amount of time.

Finally, both Wisconsin initiatives were able to secure substantial donations of time and resources and to establish partnerships that tapped into existing commercial and industrial capacity. By doing so, they were notably cost-effective. Although the total volume of products supplied in both cases was modest in terms of demand in cities like Madison and Chicago, the Field to Foodbank project and Growing Power's Carrots-to-School project deserve continued support and attention.

OBSERVATIONS ON THE POTENTIAL FOR FOOD PROCESSING AS A PATHWAY TO COMMUNITY FOOD SECURITY

With just four diverse examples to consider, definitive conclusions cannot be made about the potential for food processing to address the supply-and-demand constraints we have described. Nevertheless, some preliminary observations can be made from the brief analysis and could be tested and improved with further study.

The Relevance of Mission

One conclusion relates to the importance of organizational or program mission. The closure of the Farmer's Kitchen, for instance, might be attributed in a general sense to its ambitious but potentially conflicting goals of serving both small local farms and low-income consumers. A $7.50 lunch may be affordable, but can it sustain a restaurant that incorporates expensive ingredients? Small farms tend to seek the highest possible retail value for their produce to cover production costs. Selling to food processing outlets, restaurants, supermarkets, or public schools typically involves a much lower wholesale price. If a business tries to pay a more sustainable price to the farmer, it generally must pass that higher cost along to customers unless a continuous subsidy can cover the difference.

The mission of Homeboy Industries has always focused on the needs of its clients, so its food ventures did not aim to address demand constraints. Because Field to Foodbank was not concerned with supporting small farms, it was free to involve large producers who could make generous contributions of excess harvest. In contrast, Growing Power is a small farm that is dedicated to its own success but also to the goals of food security. Thus far it has supported both missions by significantly expanding its own scale of production, accruing subsidized land

and grants, and partnering with efficient private companies that offer significant capacity in processing, storage, and distribution.

The Need for Economies of Scale

The two Wisconsin examples demonstrate the value of economies of scale at each step in the supply chain. In terms of agricultural production, the Field to Foodbank initiative works with larger growers, whereas Growing Power rapidly expanded its carrot crop from zero to fourteen acres. After a harvest both initiatives work with processors and distributors that are large, efficient, and dependable. Both approaches result in substantial quantities of products being delivered on schedule at an acceptable price for the end user—at no cost, in the case of food bank clients. In both cases food processing is required to meet the demand, and the facilities and equipment used are capable of handling large volumes with low per-unit costs.

At the same time, both efforts represent small steps in meeting enormous needs. Second Harvest in Madison distributes more than twelve million meals per year in south-central Wisconsin. Chicago Public Schools feeds four hundred thousand students every day. If food processing is to provide an effective pathway to community food security, the two pilot efforts in Wisconsin must be scaled up considerably.

The Value of Effective Partnerships

As the two Wisconsin examples illustrate, economies of scale can be achieved when nonprofit organizations collaborate with for-profit businesses. (UW faculty and staff made important contributions in both cases.) The process of initiating and developing partnerships can be laborious. Communication across the different cultures of family farms, mission-driven community organizations, profit-driven corporations, universities, and public-sector agencies is difficult, especially when the participants are geographically dispersed. Financial concerns make these trans-actions all the more challenging because each partner faces strict limitations on what it can contribute to the cause.

The Need for Continuing Subsidies

The importance of public-private community partnerships also points to the value of charitable contributions in food processing strategies that address food insecurity. It is easier to keep the price of the final product affordable when some production costs are subsidized. For example, canned fruits and vegetables are supplied for free to Second Harvest in Madison because so much of the activity throughout the supply chain involved donated time and resources. In the Carrots-to-Schools project, each partner in the process received market prices for their products and services. However, Growing Power's ability to sell carrots at a competitive wholesale price was possible because Sysco provided rent-free land while other grants and donations supported Growing Power's overall operation. Recent public policies have also incentivized farm-to-school initiatives, in addition to the ongoing federal subsidies to the Chicago Public Schools for its school lunch program.

It is appropriate to ask how sustainable these subsidies are and whether it is possible to achieve community food security through strategies that require continuous external support. Perhaps if these strategies can achieve greater economies of scale through innovative cross-sector partnership, they will become less reliant on subsidized land, labor, and capital.

If we accept, however, that food insecurity is essentially the result of persistent poverty, subsidization in some form by state and federal governments may be necessary until poverty itself is eradicated. The existence since 1961 of the federal government's food stamp program (SNAP) illustrates how long a subsidy can continue. For Growing Power and other urban and peri-urban nonprofit farms, it may be worth considering whether the free lease of farmland, through either corporate generosity or city parks and other public land, may be more effective than SNAP dollars long-term in empowering people to achieve food security within their communities.

The Limits of Local Food

Californians are fortunate to enjoy an abundant, year-round supply of fruits and vegetables. Despite its northern climate, Wisconsin is similarly endowed with a substantial vegetable production and processing industry. Even in these states, however, there are limitations on the effectiveness of using locally grown produce to address food insecurity. For example, even California can experience shortages— especially in the face of drought. In addition, factors besides climate and availability determine where and to whom food is sold and distributed. When a product is in short supply, it is allocated to higher-paying markets, at least in the absence of subsidies or programs that alter conventional market dynamics.

Again, the importance of mission may be worth consideration. If a community's highest priority is to promote local food, its people must adapt their diet to the food that is available or else accept higher prices for produce grown under costly practices that extend the normal growing season. If the highest priority is food security, however, it may always be necessary to blend local and imported fruits and vegetables to produce healthy processed foods, with the variable cost of ingredients determining the best combinations. As local farmers expand acreage, lower their costs of production, and improve the efficiencies of cold-weather production practices, the ratio of local to imported ingredients can be increased.

The Opportunity to Enhance Flavor

The study findings cited earlier in this chapter revealed that taste may be a significant factor in why so few Americans consume enough fruits and vegetables to maintain good health. Efforts to expand individual palates through taste-tests of fresh produce, farm tours, or gardening activities may have positive effects. Cooking demonstrations can also help people incorporate fruits and vegetables into their diet. Time constraints persist, however, and some people may never develop an interest in cooking.

Although the Field to Foodbank project developed canned products for food bank clients, and Growing Power and its partners delivered freshly cut carrot coins to public school children, we can assume that both products still met some resistance. After all the effort, how many precious carrots ended up in the trash?

The abilities of the chefs who designed the menus for the Farmer's Kitchen and Homegirl Café may suggest another strategy altogether. Culinary talent can make healthy food more palatable, and despite whatever stigma may be attached to the notion of processed foods, there is no reason that determined food processing outlets cannot enhance the flavor of locally grown fruits and vegetables without degrading nutritional value.

Cost clearly remains a major constraint for consumers living in poverty. Nevertheless, can we imagine innovative strategies that combine local farms, economies of scale, effective partnerships, and targeted subsidies to create flavorful processed meals that are dense in fruits and vegetables and affordable to low-income consumers? In other words, can food processing provide a pathway to community food security?

CHAPTER SIX

Markets and Food Distribution

GREG LAWLESS AND ALFONSO MORALES

In this chapter Greg Lawless and Alfonso Morales describe the role that local markets used to play in getting fresh fruits and vegetables to consumers, and they explore some newly feasible forms of marketing and roles for marketers—the marketing component of the food system supply chain.

In order for community and regional food systems to transform and support family farms, public health, food security, resource stewardship, youth development, job creation, racial and economic justice, or any other social goals, we must understand and succeed in the marketplace. However, the marketplace is a complex concept; it involves social, political, and economic processes that unfold over days, weeks, and years. Understanding these processes can be particularly challenging in large urban centers, given the multitude and diversity of buyers and sellers; the intensity of competition; ever-changing consumer trends; the role of local, state, and federal governments; and the activities of schools, nonprofit organizations, and other dynamic forces that affect food systems.

This chapter describes the reciprocal relationship between the producers and suppliers of healthy food, on the one hand, and consumers, on the other. This relationship includes what might be called an ecology of food access opportunities, in which people and their food suppliers are in constant interaction with each other. Instead of focusing on one type of market or one mode of business organization, we emphasize the need to know the local context and leverage the assets of that context to produce viable options for purchasing healthy food. We will discuss this complex economic context after a brief introduction of historical considerations.

The chapter then describes the variety of ways that healthy food is supplied and concludes with sections that discuss the demand for healthy food and the role of extension service agents in supporting these interactions.

HISTORICAL CONSIDERATIONS

In the United States during the early twentieth century, political and social conflict along with rapidly growing cities gave rise to legal changes, infrastructural changes, and new bureaucratic forms that fashioned both the contemporary system of markets and marketing as a discipline. The variety of methods that people have used to acquire food, then and today, were born from a cauldron of competing interests; as those interests continue to grow and change, Americans have more ways to access food.

Several accounts of food distribution document practices that have emerged and evolved throughout the centuries. Scholars have described the functional interconnections and development purposes that marketplaces serve in these systems. In Europe and the United States, marketplaces played an important role in legal and economic history. The rule of law served to create markets, and the markets integrated distinct ethnic, cultural, and political groups. Many books described the development of domestic markets (Sullivan 1913; Goodwin 1929) and their role in feeding cities (Hedden 1929), and government reports described their influence (Chicago Municipal Markets Commission 1914).

Marketplaces were significant in communities around the country. A 1918 census of US public markets showed that 128 cities with populations above thirty thousand contained 237 municipal markets, including 17,578 open-air stands and 7,512 stands enclosed by market structures. The total value of market property in 1918 was estimated at $28 million (Sherman 1937).

Contemporary research on marketplaces includes histories by Spitzer and Baum (1995) and Tangires (2003). The importance of markets to immigrants, through economic opportunity and social integration, is also well documented (Eastwood 1991; Morales 1993; Eshel and Schatz 2004; Otto 2016). Markets influenced commercial redevelopment in the 1960s (Sweet 1961). They have important influences on food security (Donofrio 2007; Baics 2009) and are signficant in economics and employment (Morales, Balkin, and Persky 1995; Morales 2000), in gender relationships (Morales 2009b), in property relationships and social organization (Morales 2009a), and in social justice (Morales 2011a, 2012; Roubal and Morales

2016). One can see that a rich history is associated with marketplaces—a history inclusive of race, class, sex, law, and regulation and of activities bending toward social justice.

SOME ECONOMIC CONTEXT: THE SUPPLY
OF MARKETPLACES AND MARKETING

Previously straightforward relationships between producers and consumers became more complex in the early twentieth century. Today there are a variety of approaches to the supply of food. Food might be purchased directly from a farmer, a "pick your own" operation, a CSA farm, or a roadside farm stand. Food might also be supplied less directly through some intermediated source—for instance, through a website that offers consumers a menu of food items delivered directly to their doors by a service that aggregates food produced by local farmers. Of course, the most common intermediated sources are grocery stores and corner stores. Farmers' markets are another common source, but not all of them are producers only—many supply food that is already processed and ready for consumption, like bread, jam, jerky, and cheese. Farms also supply food directly to schools, hospitals, and other organizations. Food innovation districts are another new way to supply food chains (Day-Farnsworth and Morales 2011).

Consider how consumer spending might increase the number and perhaps the type of direct marketing outlets in major urban areas. In the seven-county metropolitan Chicago area, for example, if 1 percent of the 3.1 million households purchased a CSA share of fresh produce at $500 apiece, it would generate $15.7 million in direct payments to area farmers—enough to provide sixty-three family farms more than $250,000 apiece in gross revenue.

In terms of meat consumption, if 1 percent of Chicago-area consumers purchased 6.3 million pounds of ground and cut beef each year directly from local producers (an amount based on Midwest per-capita consumption) at an average price of $4 per pound, they would generate $25.2 million in sales and support 101 family farms (Chicago Metropolitan Agency for Planning n.d.; Davis and Lin 2005). Consumers might also spend more money at farmers' markets or on poultry. Our point is simple: where consumers spend their money influences the organization of retail opportunities.

In terms of indirect marketing, cities offer farmers many intermediated market opportunities. Farmers can sell products to consumers through restaurants, hotels, schools, hospitals, food processing outlets, supermarkets, and regional distribution outlets. In fact, a 2011 USDA report claims that 56 percent of the $4.8 billion in local food sales in 2008 was generated by farmers using only intermediated channels (Low and Vogel 2011; Martinez et al. 2010).

Farmers face two major challenges in selling through intermediated channels: (1) lowering their costs of production to achieve profitability at wholesale price levels, and (2) aggregating sufficient quantities to meet demand. The first might be achieved through economies of scale, crop specialization, mechanization, and other strategies that fall under the production component of the CRFS framework. The second might be achieved through cooperatives and relationships with private distributors, which fall under the transportation and distribution component of the CRFS framework (Day-Farnsworth and Morales 2011).

In short, the opportunity for farmers to expand direct and intermediated marketing in cities appears to be substantial. Using the seven-county Chicago area again as an example, 8.6 million residents spent an estimated $34.7 billion on food in 2010, according to the USDA Economic Research Service ("Food Expenditures"). One elusive but critical measure for tracking the progress of Chicago's community and regional food systems is the actual proportion of the market that local and regional farmers can supply over time.

THE SUPPLY OF HEALTHY FOOD

Local, state, and federal government agencies play many roles in supplying healthy food. For instance, various government entities grade food for size, shape, and nutritional value. Other agencies protect the safety of the food supplied to markets, protect the land for production, or subsidize the production of some types of food. In each of these activities, government mandates intersect with consumer demands and needs to create new opportunities in the production of a healthy food supply.

Producers then respond to public agency signals and consumer demand. They weigh the efficient production of food against production practices that might supply healthier food or that might be healthier for the farm environment. For many farmers this is a false choice; appropriate production practices can integrate efficient *and* effective production of healthy food.

The following sections explore other aspects of the complex system of supply and demand. Of course, besides being related to demand, the supply of healthy food is contingent on local regulations, local interest in producing food, soil conditions, land availability, and the vagaries of weather and climate. The discussion does not address these topics in detail, but each has a place in terms of innovative models for access to healthy food.

Innovations Related to the Availability of Healthy Food Options

On average, Americans split their food budgets approximately fifty-fifty between eating at home and eating out. Of the food eaten at home, 85 percent is purchased either in traditional supermarkets (67 percent) or at warehouse clubs and superstores (18 percent) like Sam's Club and Wal-Mart, according to the USDA Economic Research Service ("Retailing and Wholesaling"). The other 15 percent comes from many sources, including farmers' markets, CSA programs, local cooperatives and food buying clubs, mobile markets, and convenience stores (e.g., gas station markets, corner stores, bodegas, and party stores). All these vendors sell food, but they are motivated by different values and work under different legal constraints. For instance, a warehouse club represents the organizational models and values of a business, as does a farmers' market, but it differs in scale from most businesses and with respect to the laws that constrain it. For instance, a warehouse club must subscribe to employment laws that market vendors do not have to follow.

The popular outlets—along with the convenience, selection, and price competition they offer and the values they represent—are not available in many urban communities. For instance, until 2013 Detroit did not have a single national supermarket chain operating in the area. Limited choices are particularly severe in parts of the city like Northend-Central, where 34 percent of households do not have access to a motor vehicle. According to a report from Wayne State University (Galster and Raleigh 2011, 9), "in a place with a poor, fragmented public bus system like Detroit, lack of a vehicle raises huge barriers."

Fortunately this condition is changing in some parts of the city. In 2013 a Whole Foods Market opened in a relatively affluent Midtown area near the Wayne State campus; the accompanying controversy over food justice issues (Skid 2011, McMillan 2014) represented an intersection of values, convenience, and selection. In 2015 the Michigan-based Meijer grocery chain opened its second superstore within Detroit city limits (Helms 2015). Even so, wide sections of the city continue to lack supermarkets. This absence of convenience and selection is a problem of values: the right to food versus business-oriented values and the constraints of profit and law.

Fortunately, innovative strategies in three midwestern cities—Milwaukee, Chicago, and Detroit—are increasing the availability of healthy food. These efforts can be loosely characterized as alternative grassroots approaches led by nonprofit community organizations, instead of commercial pathways involving for-profit companies and traditional market channels. In short, these innovations represent a distinct mix of business and social values. Some of the most promising—and controversial—innovations involve elements of both.

One means of increasing the availability of healthy food is to start and support farmers' markets in low-income communities. For example, the Food System Assessment Study of Milwaukee in 1997 convinced the emergency food provider Hunger Task Force to make substantial investments in an underutilized outdoor market that had operated since the 1920s. Today the remodeled Fondy Food Market is open four days a week during the growing season; several dozen predominantly Hmong vendors fill an area of thirty-eight thousand square feet to provide fresh produce to African American residents on Milwaukee's north side. The market is now operated independently of Hunger Task Force.

In Chicago the nonprofit Experimental Station established the 61st Street Farmers Market between the University of Chicago campus in Hyde Park and the lower-income Woodlawn neighborhood. By locating adjacent to an upper-income community, the market generates enough revenue to attract vendors whose produce might not otherwise be available to Woodlawn residents. The market accepts food stamps, which ensures lower-income consumers access to the healthy produce. Markets like these supply locally sourced food, which reduces transportation-related greenhouse gas emissions from imported food. Mitigating climate change would certainly be a positive benefit of such markets (Bentley and Barker 2005).

During Michigan's growing season, Detroit Eastern Market attracts as many as forty thousand customers every Saturday to an area northeast of downtown that is accessible to the residents of disadvantaged East Side neighborhoods. The nonprofit Eastern Market Corporation also partners with fourteen Detroit-based community groups to support farm stands in eleven locations where healthy and affordable food would otherwise be inaccessible to the residents.

Besides markets, there are other modalities for enhancing food access. Another direct marketing innovation involves produce trucks that run routes through low-income communities, much like traveling ice cream vendors. Peaches & Greens Mobile Market, one of three produce-truck operations in Detroit, has reportedly been successful in part because its parent organization also runs a retail market, providing a secondary outlet for perishable produce (Central Detroit Christian n.d.). Jurisdictions regulate these distribution activities to address legitimate public health concerns, but regulatory barriers can also impede innovation in the field. However, good arguments exist for experimentation and the moderation of such problems (Morales and Kettles 2009a).

Efforts to improve food availability through more traditional retail channels also merit attention. In communities that lack supermarkets, residents often rely on corner stores and other small outlets for their groceries. In the large West Englewood neighborhood of Chicago, for instance, where only one independent grocery store offers fresh produce, residents are served by twelve corner stores and eleven "food and liquor stores," where fresh produce and other healthy food options are often overpriced, of poor quality, or lacking altogether (Chicago Policy Research Team 2010). In Detroit, a project led by Wayne State University introduced or expanded fresh produce in eighteen corner stores in 2010 (Pothukuchi 2011).

Finally, a number of efforts are under way in Detroit, Chicago, and Milwaukee for independent supermarkets, large corporate chain stores, and mainstream distributors to expand their availability of healthy food. This may represent a new configuration of business and social values. Recent efforts involving large corporate food companies have also generated considerable controversy in food security circles over unfair labor practices, local control, and an inability to account for loss of capital (Patel 2008; Morales 2011b; Goodman, Du Puis, and Goodman 2013). These issues are addressed under other components of the CRFS framework. Regarding markets and marketing, the influence of corporate involvement can be assessed over time in terms of its effect on the availability, affordability, and appeal of healthy foods in low-income communities. Whether these efforts will be successful, and how success is measured, will require significant new research.

Networks and Collaborative Marketing Partnerships Drive Food Hub Development in Cedar Rapids

JASON GRIMM

The vision of Iowa Valley Resource Conservation & Development (IVRCD) for East Central Iowa is of a sustainable and prosperous regional community operating in harmony with its cultural and environmental resources. We have learned that farmers' markets and other direct outlets are not enough to make their regional food and farm businesses economically viable. Financial viability requires working in multiple distribution modalities. Here I describe how these businesses increased revenue by developing the market for local food.

IVRCD believes that the key to market development is through fostering networks between farms and buyers and establishing collaborative marketing partnerships between farms. For instance, the regional supermarket chain Hy-Vee built a relationship with the Iowa Valley Food Co-op, a direct-to-consumer cooperative of producer and consumer members in Cedar Rapids and Iowa City. The co-op coordinated local produce orders for seven Hy-Vee stores from nine of the co-op's member farms.

In 2014 an IVRCD contribution made the co-op's partnership with Hy-Vee a win for its member farms. By forming a partnership with the regional food bank Hawkeye Area Cooperative Action Program (HACAP), IVRCD established a regional distribution hub for the co-op's sales to the Cedar Rapids Hy-Vee stores. Co-op producers aggregated their individual Hy-Vee orders into twice-weekly deliveries to HACAP. There the aggregated orders were palletized and delivered to the Hy-Vee stores by HACAP drivers making regular runs to pick up "rescued food" made available by Hy-Vee to the food bank. From May through September 2014 the co-op producers sold eighty-three different types of produce to seven Hy-Vee stores in 3,278 cases, resulting in more than $80,000 in sales at an average case price of $24.16.

The collaborative marketing that the co-op's member farms established provided more than a place to sell their produce. The collaboration was an opportunity to leverage multiple marketing advantages at once. The member farms sourced wax produce boxes, twist ties, and plastic box liners and even developed their own price look-up stickers in collaboration with one another. The stickers labeled all the produce under the Iowa Valley Food Co-op brand while also showcasing each farm's identity. In the spring of 2015 this collaboration went even further by sourcing three tons of seed potatoes and spreading the cost among the member farms.

The Iowa Valley Food Co-op and its member farms are continually developing new ways to leverage their network and improve the economic viability of the individual farms. IVRCD is helping by fostering networks that leverage assets across community food system sectors, thus creating a rising tide that lifts all boats.

Innovations Related to the Affordability of Healthy Food Options

Efficiencies of scale, buying power, and fierce competition explain why large food distributors and corporate chain stores are able to offer high-quality food, wide selection, and low prices—at least in communities where they can generate profit.

It would be useful to track the sources, quantities, and prices of locally and regionally grown foods that make their way into traditional supermarkets in low-income communities. Of course, price is a major concern for most consumers—especially in places like Milwaukee, where in 2010 29.5 percent of the residents and 46.1 percent of the children lived below the poverty level, an annual income of $22,314 for a family of four (Sears 2011).

A 2011 food retail study by a neighborhood news reporter in Milwaukee showed the substantial range in prices in nine area food stores for a cart of twenty-two grocery items (McGowan 2011). The report also demonstrated how community members can effectively gather legitimate and valuable market data about their community and regional food systems. The collection of market data is a critical step in understanding local food systems. People have questioned the merits of partnering with nonlocal corporations to improve those systems, for both ideological and substantive reasons. However, an important test of such collaborations is determining their actual effect on food availability and affordability in low-income communities, and those tests will require some systematic data analysis.

Besides studying retail opportunities, a complete assessment of community and regional food systems would include the role that charitable donations play in the local food supply chain. Free and discounted foods are an obvious way to address affordability, and efforts to improve their nutritional quality with locally grown produce must be measured and evaluated (Ramde 2011). Another way that food is made more affordable involves federal assistance programs like SNAP, WIC, and the National School Lunch Program.

In Chicago the 61st Street Farmers Market has enabled SNAP recipients to double their benefits for up to ten dollars spent on fresh produce at the market. The market's sponsor has carefully recorded data about the program's participation rates and effect in the hope of receiving continued support from state and city governments (Experimental Station n.d.). A similar program, called Mo' Bucks, was introduced in 2009 by Detroit's Eastern Market Corporation; the program involves four farmers' markets and the Peaches & Greens Mobile Market (Fogelman

2009). Such programs are common and are increasingly the subject of evaluation and support (Dundore and Morales 2015).

National programs play an integral role in local food marketplaces by affecting the purchasing power of consumers. National efforts on behalf of local and regional food systems were recently supported by the 2014 Farm Bill, which provided $30 million in mandatory annual funding, tripling the funding allocated in the 2008 bill. The bill also expanded—and renamed—the Farmers Market Promotion Program, which is now the Farmers Market and Local Food Promotion Program. This reconstructed program supports direct farm-to-consumer marketing such as farmers' markets and community-supported agriculture, and it provides grants to enhance farm-to-institution networks (e.g., schools and hospitals), establish or improve food hubs, and support other local and regional food businesses. Community and regional food advocates hope that the program will materialize support for developing and practicing modes of food aggregation and distribution that address social and economic needs.

The farmers' market Metrics and Indicators for Impact program was initiated in 2014 with a USDA grant to UW–Madison. The activities of this program help farmers' market managers understand the social, economic, ecological, and health impacts of their markets on the community. Understanding impact is a vital aspect of food system activities in general and for markets in particular in this instance because it improves manager decision making and communication with consumers and other stakeholders. Understanding impact is also important to the USDA and other grant makers, who want to know the impact of their investments.

Understanding the Appeal of Healthy Food Options

Getting healthy foods into low-income communities at affordable prices is a major accomplishment. However, there is no guarantee that people will buy and consume that food. An important question for community and regional food systems is whether local foods can be effectively marketed to make healthy eating more appealing to low-income populations.

While literature exists on marketing techniques to promote unhealthy food in these communities, far less is written about proven marketing strategies that actually increase sales of healthy foods to low-income consumers—especially to people of color (Grier and Kumanyika 2008). The business literature on ethnic and multicultural marketing could be instructive if it were applied to local and healthy food marketing (Tharp 2001; Miller and Kemp 2006; Korzenny and Korzenny 2005).

A basic precept of marketing is "Know thy customer." Research helps in achieving this goal. Secondary market research uses data gathered by government agencies, industry trade associations, and other external sources to better understand a particular population. Data always require interpretation, and people can interpret data in different ways. For instance, 2010 census data showed that a neighborhood north of Gratiot Avenue on Detroit's East Side (census tract 5161) was home to 734 people in a one-half square-mile area. Of these residents, 97 percent were African American, and 40 percent had earned a high school diploma. About 88 percent of the households earned less than $35,000. One interpretation of this data is that the neighborhood is disproportionately poor and not well educated, but this data does not explain whether the neighborhood has access to healthy food or if the people can afford the food found nearby.

Secondary data can also provide insight into the behavior and attitudes of potential customers. For instance, a USDA report on how much time Americans spend on food (Hamrick et al. 2011) is replete with information about grocery shopping, meal preparation, and eating habits in the United States. Among its many insights is that people living below 185 percent of the poverty level spend significantly more time per day (40.3 minutes) preparing meals than people living above this threshold (29.5 minutes). Applying this information to the Detroit case would require more analysis to make some inferences about the behavioral patterns of census tract 5161's residents.

By carefully combining multiple sources of secondary data, one can draw preliminary conclusions about potential customers in a certain population. Using the preceding data, for example, one might conclude that selling fresh ingredients with recipes would be a more effective marketing strategy in census tract 5161 than in more affluent areas of Detroit, where people spend less time preparing meals.

Primary market research is another mode of knowing your customer. Primary research applies questionnaires and interviews to a sample of a given population and can provide more complete and reliable information than secondary data. For example, a survey of census tract 5161 residents might contradict the conclusion in the previous paragraph. It could also provide richer and more detailed information, such as the types of fresh ingredients the residents want, where they want to purchase them, and how much they are willing to pay.

Primary research can provide qualitative data about consumers' attitudes or opinions. For instance, a market environment or promotion that appeals to one segment of the population may have little effect—or the opposite effect—on another segment. A survey might find that the notion of "connecting to the land" or "knowing where your food comes from" motivates suburban or college-educated consumers to join a CSA farm, but it may not have the same appeal to immigrants who are seeking to escape low-paying farm labor.

Grassroots marketing efforts are under way in cities like Chicago, Milwaukee, and Detroit to enable and persuade disadvantaged populations to eat healthier foods. These efforts include the four P's of marketing: product, price, place, and promotion. We have discussed locally grown fresh produce (product), double-value SNAP benefits (price), and farmers' markets and mobile trucks (place). Branding efforts like Grown in Detroit and Grown in Michigan tap into local pride (promotion). Because measurement is difficult and costly, the extent to which these efforts are affecting purchasing behavior and health outcomes is unclear. One study (McCormack et al. 2010) examined whether farmers' markets and community gardens are accomplishing their goals in this area; the results were encouraging but inconclusive.

Assistance with Markets and Marketing from Local Extension Offices

GREG LAWLESS

At a symposium in Kansas City in May 2011, members of the North Central Cooperative Extension Association (NCCEA) identified a need for more research and outreach in what they called metropolitan food systems. A draft executive summary from that event mentioned the need to improve "food literacy" so that all metropolitan consumers can "make informed choices about food access, nutrition, preparation, and better health generally."

In its follow-up to the symposium, NCCEA requested guidance for how university extension systems can do more to support metropolitan food systems. The ten themes they developed to address the topic closely reflect the components of the CRFS framework.

There are four programmatic areas in which extension systems might support markets and marketing in community and regional food systems, or what the NCCEA has called "consumer-centric food information and marketing":

- Local market data collection and analysis
- Ethnic and multicultural marketing
- Nutritional evaluation of market environments
- Entrepreneurship education

A wealth of free and affordable information is available to describe local populations and market conditions. Data from the US Census Bureau, the USDA's Economic Research Service, the Bureau of Labor Statistics, and the North American Industry Classification System are especially rich and valuable. However, it takes time to become familiar with these resources and to develop the necessary skills and tools to compile and analyze data. A local extension office can help by providing data, software, and analytical frameworks (UWEX n.d.). The Center for Community Economic Development at the University of Wisconsin Extension has done exemplary work compiling community economic profiles and indicators to measure community food systems.

There is a growing body of academic and business literature on the topic of ethnic and multicultural marketing. The subject raises difficult issues about cultural differences, stereotypes, and racism. Facilitated discussions might be necessary between the practitioners who are promoting healthy food choices and the communities they hope to affect. Results from discussions and observations should be compiled and summarized for general audiences and food vendors. Again, a local extension office can help (Tharp 2001; Guion and Kent 2005).

As more people become aware of the connections among health, nutrition, and food systems, methods have been developed to measure and assess the nutritional status of different food environments. The Nutrition Environment Measures Survey (NEMS) first developed by Emory University in Atlanta, is a frequently cited example (McKinnon et al., 2009). It can be used to evaluate restaurants, supermarkets, and even vending machines. However, like market data sources, useful tools such as NEMS are time-consuming to learn and implement. Iowa State University Extension supports a website that provides extensive resources on using NEMS to evaluate the workplace vending-machine environment.

Both nonprofit organizations and for-profit companies that operate in local food systems need to develop and implement marketing strategies that fit within their broader business plans. Continual research and planning are critical to sustaining success in the marketplace, but it is difficult to make the time and use it effectively.

SUMMARY

This chapter described the business of markets and marketing from two nonbusiness perspectives: academic and consumer. The approach was to ask questions of interest to business that focus on the consumer and to point out organizational innovations that increase the number of ways consumers can access healthy food. In summary, we will describe the five C's that constitute important elements of markets and marketing: cooperation, customer orientation, cost, community, and contact.

The first C, cooperation, is what establishes farmers' markets, food hubs, and the many alternative approaches to stores that help people access healthy food. Cooperation can take many forms; cooperatives themselves are a renascent business organizational model. Existing businesses can form partnerships of various kinds for achieving goals, such as partnerships that include affordable access to healthier food.

The second and third C's are customer orientation and cost. Marketing actions attempt to convince consumers about the merits of a product; in contrast, public health programs emphasize the value of healthy food to individual health, and planning programs emphasize the importance of social *and* economic values. Innovations in food marketing will include various mixes of these distinct values and will be organized in various ways.

Community, the fourth C, grows within its social, economic, and political context, shaped by the various values of those in the community. If cost considerations constantly force people to accept less costly and often unhealthy alternatives, then how might the community respond? How will business, law and policy, and other people react to this quandary?

The range of influences that people and organizations have on each other is the fifth C, contact. Marketing typically assumes that consumers are actors, and it attempts to influence their actions and behavior to get them to purchase products. However, consumers also interact with others and their environment (Morales 2002). Consumers are complex; they play multiple roles and participate in society through interaction with others in their various consumer activities. For instance, consumers visit farmers' markets and might make purchases, but they might also pause from a busy day for a moment to take in the pleasant setting, interact with others in that environment, pick up some information, and observe a cooking demonstration or even a political demonstration in the same setting.

TABLE 5. Challenges and Innovations in Community and Regional Food Distribution: Increasing Market Access and Farmer Profitability

Challenges	Innovations
Lower profits for products sold anonymously	Showcasing of product origin through knowledgeable sales representatives, packaging, and point-of-sale merchandising
Need for improved delivery coordination and physical infrastructure (e.g., temperature-controlled storage and vehicles)	Development of formal and informal food hubs and equipment or facility sharing through farmer-distributor networking opportunities and public-private partnerships
Need for scale-appropriate product-tracking technologies	Quick response, RFID (radio frequency identification), and UPCs (universal product codes) to foster traceability, although their cost, information storage capacity, and intended uses vary
Lack of knowledge about actual cost of distribution	Development and dissemination of cost-of-distribution workshops and calculators to help farmers and small-scale distributors determine if and when outsourcing is profitable
Inconsistencies in interstate transport regulations	Cooperation with research, trade, and advocacy organizations to harmonize interstate transportation regulations, which will improve compliance and foster interstate trade
Unreliable local supply	Development of farmer training and resources for preseason production planning, wholesale market production, postharvest handling, pack size, and pricing practices

In other words, people are not just consumers; they seek a variety of interactions and contact that might have little to do with purchases. Progressive food system practices should foster such contact.

Ideas for Further Development

One of the most vexing concerns in food distribution is the fair pricing dilemma—"business models that maximize farmer profits often make products too expensive for low-income consumers" (Day-Farnsworth and Morales 2011). Cost-

TABLE 6. Challenges and Innovations in Community and Regional Food Distribution: Increasing Accessibility and Affordability of Healthy Food in Small Retail Stores

Challenges	Innovations
Difficulty meeting minimum purchase volumes of large suppliers	Leveraging of the collective buying power of corner stores through consortium purchasing
Need for SNAP- and WIC-approved vendor status	Assistance to corner stores in achieving SNAP- and WIC-approved vendor status and acquiring EBT machines
Difficulty financing infrastructure (e.g., cooler storage and display cases)	Financing of loans and grants to leverage private investment for infrastructural improvements that increase stores' capacity to store and sell fresh products
Store owners' lack of experience in marketing and merchandising fresh products	Improvement of marketing and merchandising skills through consultation, free promotional materials, and fact sheets on stocking and display tips
Uncertain market demand	Engagement of store owners and local organizations in neighborhood market research to determine which fresh products are in demand. Improvement of nutrition literacy and the appeal of healthy products with prominent, well-lit displays, store facade enhancements, and regular product culling

saving innovations and improved efficiency in scale-appropriate transportation and supply chain management can potentially improve farmers' access to local and regional markets and translate to more affordable consumer prices, but further research is necessary to learn the following:

- How can accurate measurements be made to quantify changes in the supply and sale of local foods in a complex metropolitan marketplace?
- What are the consequences of partnering with national food corporations to improve community and regional food systems, and how can these consequences be tracked and evaluated?

- Are the marketing strategies to address food availability, affordability, and appeal in minority communities culturally appropriate and effective?

Several challenges and innovations are listed in tables 5 (Day-Farnsworth and Nelson 2012; Bittner et al. 2011) and 6.

The authors sincerely acknowledge the generous contributions of Lindsey Day-Farnsworth and Amanda Hoffman to this chapter.

CHAPTER SEVEN

The Consumer

Passion, Knowledge, and Skills

MONICA THEIS

In this chapter Monica Theis presents two recent examples of her outreach work in order to deepen the understanding of the concept of food literacy. She argues that consumers must be an important part of the food supply chain and notes that this will require some effort on the part of consumers and food system advocates. This chapter covers the preparation and consumption components of our food system supply chain.

For nearly a decade now, food activists—including journalists, academics, chefs, politicians, and celebrities—have relentlessly advocated for improvements in the American food culture. Their messages can be condensed to five core tenets:

- Eat real and healthy food.
- Eat less.
- Emphasize plants over animal sources.
- Buy locally, and preferably organic.
- Prepare all food from scratch in your home.

Michael Pollan and Mark Bittman have been particularly persistent with these messages. For example, in his book *In Defense of Food* Pollan (2008, 1) states, "Eat food, not too much, mostly plants." Bittman (2015) takes these messages to the next level with cookbooks and media appearances, emphasizing the implied simplicity and cost savings of cooking for oneself. His best-selling books include *How to Cook Everything* and *How to Cook Everything Fast*. However, absent from these works is a full and honest explanation of how to get good food from its local source to the table while keeping in mind consumer values and modern household realities.

Although these messages are well-intentioned and commendable in the context of health and environmental stewardship, there is a disconnect between food activists' ideals and what it actually takes to make these ideals a reality in the vast majority of American households and institutional kitchens. Wild assumptions are being made by food activists and food enthusiasts about a passion for cooking, a knowledge of food, cooking skills, facilities and equipment for storing and preparing food, and available resources, including time and money. Many food activists lack the knowledge, appreciation, and respect for what is required to get food to the table in a manner compatible with the reality of day-to-day life. Numerous studies and publications have addressed this more comprehensive analysis of home cooking (Bower, Elliott, and Brenton 2014; Coveney, Begley, and Gallagos 2012).

The purpose of this chapter is to explore the complexities of the farm-to-table continuum from the perspective of various consumers. Reflective stories from community-based food education and cooking classes are used as brief case studies. Two cases in particular highlight innovative approaches to consumer education on food preparation. The first case is a food education program launched in partnership with a food pantry at the Middleton Outreach Ministry in Middleton, Wisconsin. The second is a program developed in partnership with the University of Wisconsin–Madison Police Department as part of its wellness initiative.

BACKGROUND

Various concepts of food systems, including supply chains, food justice, and food security, have been addressed earlier in this book. A comprehensive understanding of these concepts is often referred to as *food literacy* (Vidgen and Gallegos 2011). Consumers should be encouraged to become more food literate in an effort to fight chronic disease, protect the environment, and support local economies. Consumers are becoming increasingly aware of the importance of keeping these issues in mind when making food purchases. Although environmental protection, justice, food security, and support of local economies are relevant issues to many consumers, price, familiarity, and convenience are still the primary drivers of how they choose to spend their food dollars. For consumers who are willing to support a more sustainable food system, a common barrier is lack of food knowledge and preparation skills.

Chapter 1 addressed the five As of food security: available, affordable, accessible, appropriate, and acceptable. However, these concepts only get food in the door; they do not get it on the table. Therefore, another A must be considered: *appealing*. Consumers want their meals to be visually appealing and appetizing.

It is important to recognize that cooking in the home has become a much rarer phenomenon in the last several decades. The large-scale movement of women into the workforce, the introduction of inexpensive processed foods, and our hyperbusy lifestyles have all contributed to this dramatic social shift. Similarly, the ability to cook from scratch in institutions has also faded away, primarily because of the economic hardships of getting fresh food and the skilled labor to prepare it. As consumers, governments, and communities come to realize the long-term health and environmental implications of this problem, more and more people are working to bring fresh and local foods into their homes and community kitchens. This is the action that activists like Bittman and Pollan are trying to inspire. It stands to reason that food knowledge and the cooking skills required to successfully transform fresh food into an enticing meal also need to be reintroduced. For the purposes of this chapter, cooking literacy encompasses the ability to do the following:

- Identify food in its raw form by sight.
- Select seasonal, local food at the point of purchase for peak quality.
- Store food properly to maximize safety and retain quality.
- Prepare food using methods that minimize waste and retain natural flavors, nutrients, and color.

In terms of cooking literacy, consumers raise the following practical and basic questions, especially regarding fresh produce:

- *What is that?* Many consumers have never seen vegetables in their raw forms, much less learned how to prepare them.
- *What should I look for when buying fresh produce?* Many consumers do not know how to select fresh produce and are unfamiliar with the concept of seasonality.
- *How do I store it in my home, and how long will it keep?* In addition to being sensitive to the cost of fresh produce, consumers also fear wasting food that may spoil before they can prepare it.

- *How do I prepare it, and what kitchen equipment do I need?* Many consumers grew up in households where cooking was not part of the daily routine. As a result, they may not own or even recognize basic kitchen tools. In addition, home and institutional cooks alike may be unfamiliar with cooking methods that maximize yield and flavor.
- *How can I still enjoy my favorites?* Consumers are not willing to forgo their favorite and familiar foods, especially those that reflect a cultural tradition.
- *How do I make it taste good?* This is very important, since many consumers have grown up on food that is full of salt, sugar, and fat; the natural flavors of foods in their raw forms simply do not appeal to their abused palates. Consumers are also unfamiliar with the subtle flavorings of many fresh herbs and spices.

DESIGNING EDUCATION PROGRAMS FOR COOKING LITERACY

Many people, especially those of us in academia, have suggested that education is essential in making the transition from "heat and serve" meals to thoughtful food selection and preparation from scratch. Yet even though education is certainly important, knowledge of food, food systems, and their broad economic and social implications is not enough (Mancino and Kinsey 2008; Schaeffer and Miller 2012; Worsley et al. 2014). Many people will bluntly state, "I know what I am *supposed* to eat; what I need is guidance on how to prepare it with the time I have." Thus, community and institutional education programs must be developed that cater to the specific needs, values, and available resources of targeted audiences. These programs should include skill development that is realistic in the context of a person's passion, interests, and available resources.

General principles of instruction suggest that the learning needs and knowledge gaps of an audience should be identified before effective program design can begin. For food and cooking literacy, these preliminary assessments, if they are done at all, typically focus on food knowledge and cooking methods. Assessment of a person's passion for cooking is rarely considered. Food enthusiasts often wrongly assume that consumers will enjoy cooking if they just have the knowledge and skills to transform raw products into good, edible fare. Program design that acknowledges and accommodates passion and personal values has more potential to ensure engagement and promote meaningful, sustainable change.

In 2014 and 2015, opportunities to apply innovative approaches to cooking literacy emerged in the Madison area. A local food pantry and the UW–Madison Police Department separately approached the UW-Madison Department of Food Science and the CRFS project seeking guidance on education programs on food and cooking literacy for their clients and staff, respectively. The rest of this chapter elaborates on these two examples, which showcase the importance of focusing on the target audience's values, lifestyles, and available resources.

THE FOOD PANTRY AT MIDDLETON OUTREACH MINISTRY

Middleton Outreach Ministry (MOM) is an established nonprofit organization leading a communitywide effort to prevent homelessness and end hunger throughout Middleton, western sections of Madison, and Cross Plains, Wisconsin. Through the support of business partners, area schools, faith-based communities, service organizations, and numerous individuals, MOM provides free food, clothing, housing assistance, emergency financial assistance, and special services for seniors. Several UW–Madison departments, including Food Science, Horticulture, Agronomy, Plant Pathology, and Landscape Architecture, work in partnership with the MOM Food Pantry. Two aspects of the MOM program that unite these varied disciplines are its food distribution pantry and its on-site gardens.

The Food Pantry is MOM's flagship operation. Operating six days a week year-round, the pantry serves nearly seven hundred households each month, including nearly two thousand children. The Food Pantry is recognized for its progressive and unique model. Guests of the pantry can take as much food as they can use and visit as often as they need to, effectively eliminating hunger for anyone in the service area. MOM partners with Second Harvest Foodbank (see section) and the Community Action Coalition for South Central Wisconsin and receives support from businesses and foundations throughout the community. Its services are supported by the commitment of well over two hundred volunteers. In 2014 MOM distributed more than one million pounds of food.

MOM also operates a robust garden program that includes acreage dedicated to the pantry, a teaching garden, and individual plots that MOM's clients can use to grow their own produce. In 2014 the gardens produced more than eight thousand pounds of food.

The Need to Move Food Quickly

One of MOM's core challenges is the need to move food products fast enough to prevent spoilage. This is especially true of produce from the pantry gardens and fresh

FIGURE 12. Fresh and refrigerated foods are displayed attractively and strategically by Middleton Outreach Ministry's Food Pantry to entice consumers to select them. (Courtesy of Monica Theis, CRFS)

foods from area suppliers, including fruits, vegetables, dairy products, and meat (fig. 12). In addition, large donations of shelf-stable dry goods, such as pasta and canned foods, must move quickly so that space remains available for incoming donations.

Some barriers to product movement persist, including a lack of knowledge of how to prepare food and an aversion to unfamiliar products. To better understand these problems from the perspective of its clients, MOM formed a client advisory group in 2014. The group met that spring to discuss what clients want and need to know about food and its preparation. From the preliminary discussions it became evident that clients have unique needs in terms of the facilities required to prepare food. Some clients live in housing with fully equipped kitchens, but others live in units where the only appliance may be a microwave oven.

Tasting Kick-Off

MOM decided that the first step in expanding clients' food and cooking knowledge would be to develop a tasting program that showcased the food items that are abundantly available throughout the pantry. Beginning in January 2015, students in a UW–Madison food science class called "Organization and Management of Food Service" worked with MOM's distribution manager to design a series of recipe tastings for clients who regularly visit the food pantry. The overall goal of the project was to move products off the shelves, but the following other objectives were also identified:

- Emphasize food items and ingredients that are commonly and abundantly available through the pantry.
- Showcase recipes that are easy to prepare at home with minimal time and equipment.
- Emphasize the health benefits of foods available through the pantry.
- Introduce new or unfamiliar foods to clients.

The tastings began in March 2015 in observation of National Nutrition Month. Six tastings were designed and offered over six consecutive days, March 23–28. Recipes were selected from a cookbook designed for food pantry patrons: *Good and Cheap* by Leanne Brown. The initial series showcased the following recipes:

- Monday: Half-veggie burgers
- Tuesday: Corn soup
- Wednesday: Beef stroganoff
- Thursday: Vegetable quiche
- Friday: Vegetable jambalaya
- Saturday: Pasta with tomato and eggplant

The dishes were prepared beforehand and reheated on-site using a microwave oven and a two-burner stove. The tastings table was set up near the exit of the food pantry to attract clients as they were leaving. Bags of ingredients related to the recipe of the day were available at the point of exit as well. Students and volunteers offered product samples and kept track of the following information:

- The number of men, women, and children who tasted the product
- Comments made by clients about the product, especially whether they liked the dish and would consider making it at home
- The number of ingredient bags taken

Feedback from MOM's clients, volunteers, and staff clearly indicated that this initial pilot program was successful not only in terms of meeting its objectives but also in building a deeper sense of community between the volunteers and the clients. As a result, MOM decided to formally integrate the program into the services it offers. During the summer of 2015 a student volunteer designed a protocol for use by the regular volunteers. In September 2015 a MOM volunteer took ownership of the program to offer tastings throughout the year. Fresh produce from MOM gardens is the focus during the growing season.

Postprogram assessment of learning is another aspect of community education to consider. As the project moves forward, it will be important to determine the long-term influence of the program on MOM's clients. In particular, MOM must determine if the program actually inspired clients to cook more at home. Previously published information on the effects of community education will serve as a model for developing an objective assessment tool (Barton, Wrieden, and Anderson 2011; Dunn et al. 2014).

Field to Foodbank: A Program of Second Harvest Foodbank of Southern Wisconsin

OONA MACKESEY-GREEN

During lunch at a Midwest Food Processors Association meeting in 2012, Jed Colquhoun said, "the fate of a carrot crop" sparked discussions that led to the creation of Field to Foodbank, a program of the Second Harvest Foodbank of Southern Wisconsin. Colquhoun is a faculty member in the UW–Madison Department of Horticulture and a UWEX specialist.

"On one side of me was a carrot grower," he recalled, "and on the other side was an individual who did food resourcing for Second Harvest. I asked the carrot grower how the year was. He said it was great but that they unfortunately froze a couple hundred thousand pounds of carrots in the field because they just couldn't get to them. They had so much volume and their yields were so high that they lost the crop. Of course, the food resourcing manager dropped his fork and exclaimed at the quantity of lost produce.

Crop yield is necessarily uncertain. In some years the yield falls short because of environmental events; in other years there is a bountiful harvest with a higher yield than can be used through traditional food supply chains. Rather than accept unpredictability, the three men recognized an opportunity looming between a need—food insecurity in Wisconsin—and the state's agrarian strength. Once the match was made and the conversations began, said Colquhoun, "it happened rather organically."

According to Colquhoun, the "magic" of Field to Foodbank occurred as the program's network sprawled across the state, working across logistic chains to continue matching the strengths and struggles of various partners united by a desire to deliver excess food to empty tables.

Field to Foodbank operates with large volumes of crops in order to serve the needs of the participating farmers statewide. Colquhoun's expertise in commercial fruit and vegetable production and experience supporting all scales of production, from small-scale, diversified CSA projects to massive potato farms, allowed him to provide the necessary technical assistance to farms of varied sizes and geographical locations as their overproduction was consolidated. In coordinating the harvest, transport, processing, and distribution of large volumes of produce statewide, Colquhoun explained, "we have three issues to deal with: perishability, food safety and traceability, and seasonality."

The partnerships involved in Field to Foodbank present particular challenges within these broader issues. Regarding the first issue, perishability, Colquhoun admitted, "It was a steep learning curve for me to recognize that many food pantries and some food banks don't have refrigeration. They have no capacity to handle fresh produce. The perishability component, what food bankers call the last mile, is critical. It is the linchpin in this whole program. The last thing you want to do is give somebody something they throw away a couple days later because it has gone bad." The group quickly realized that distributing large volumes of fresh produce to food banks was impractical, limiting both the shelf life and the seasonality of the products.

"The tide changed on that, really, when the food processors got involved and donated run time and processing plants, as well as cans." Colquhoun emphasized that Field to Foodbank attracted new partners through its commitment to developing consistent relationships and returning donations to local communities rather than through financial incentives. "This effort doesn't fit the traditional growth in urban agriculture, but it serves a lot of that need. Food insecurity isn't just an urban problem at all. The most common question I get from agricultural suppliers is 'How much of this food will come back to my community? Because I have neighbors in need.'" Agricultural producers of crops that are not directly consumed (like wheat, corn, and soy) can support their communities through Field to Foodbank's Invest an Acre program, which donates a portion of the commodity profits directly to local food pantries.

A partnership with Cummins Filtration allowed Colquhoun and his colleagues to develop a complex logistics chain and tracking mechanism to tackle the second central issue, food safety and traceability. Field to Foodbank uses a system based on the electronic logistics chain that Cummins Filtration applies in its industrial manufacturing plants to track input, output, and efficiency; however, it was customized for local food system networks. "We distinguished how you handled fresh product versus processed product and different logistic chains for each of those," explained Colquhoun.

Tracking the logistics chain also helps with the third issue, seasonality, the challenge that most of a crop matures around the same time of year. By knowing the schedules and contracts of packers and canners, Field to Foodbank can arrange to move crops to processing facilities when and where the capacity is available.

As new challenges arose, Colquhoun most often played matchmaker. "Field to Foodbank has been a great program," he said, "but it is really based on relationships: bringing people together in nontraditional ways and not only making them aware of local food insecurity but also connecting them in ways that are useful in terms of sourcing food. The food supply chain has become so efficient that you can have a few conversations and get a lot done. Those folks have all worked together in very agreeable ways for decades." Partnering with food pantries often meant taking field trips with representatives from the pantries to local farms to learn about the food supply chain from an agricultural standpoint. During these excursions the participants learned more about farming and built relationships.

A statewide association of food banks, Feeding Wisconsin, grew from these bolstered collaborations; through the nationwide organization Feeding America, Feeding Wisconsin supports a network that allows food banks to distribute the high volume of local crops made available through Field to Foodbank by trading products internally, according to their needs and capacities.

"Really what we recognized was [that] individual entities and growers have their strengths, but none of them have the whole system," said Colquhoun. "That's where the magic of Field to Foodbank happened, realizing that we needed to combine cross-institutional and cross-company logistic strengths in nontraditional ways. Let's say we have a field of carrots, but the grower doesn't have the capacity or the time to harvest them. Maybe a neighboring grower or agricultural entity does, and it brings its equipment over. Its harvests it, and then we call over somebody willing to volunteer some trucking capacity to truck it to a canning company that is willing to can it. And then it's stored in a warehouse that belongs to another company, and then it is distributed from there."

In the past, any missing link in this chain might have led to a carrot crop frozen in the ground while people in urban and rural communities across Wisconsin remained food-insecure. Field to Foodbank's supply chain innovations tackle those missed connections. As the organization expands and continues to work through ongoing challenges, Colquhoun hopes that these bridges will continue to increase the awareness of food insecurity and serve as a model for networks nationwide.

THE UNIVERSITY OF WISCONSIN—MADISON POLICE DEPARTMENT

During the summer of 2014 the UW–Madison Police Department (UWPD) approached the Department of Food Science to inquire about a partnership in support of its newly implemented wellness program. The UWPD had made a deliberate decision to invest in improving its human resources, with the specific goal of meeting the changing needs of the people it serves on campus and in the community. The purpose of this partnership was to develop small-group instructional sessions on nutrition strategies specific to the needs of the UWPD staff. Initial conversations with program leaders indicated that a focus on food and its preparation would be of interest and value to the staff.

Twenty officers and members of the UWPD support staff signed on for the inaugural program during the fall of 2014. Classes were taught in the Food Applications Lab of the Department of Food Science and were offered in partnership with a dietitian from the UW Athletic Department. Facility use fees were waived with the understanding that this first attempt to offer a food and cooking program would assess how the UWPD could become part of the educational programming at UW–Madison. In other words, both departments were seeking the freedom to explore innovations in education and outreach.

The twenty participants ranged in age from twenty-two to sixty-five and were evenly split between men and women. Several participants were parents interested

TABLE 7. Food Knowledge and Cooking Skills Survey

Food for Health Wellness Program
UW–Madison Police Department, in partnership with the UW–Madison Department
of Food Science

Name (optional):_____

List or describe any special dietary needs that you want us to accommodate during the Food for Health classes (e.g., vegetarian, food allergies, gluten-free).

Which of the following best reflects your *passion* for cooking? (Check one.)

_____ I love to cook.

_____ It's OK; I'll do it if I have to.

_____ I don't like to cook.

Which of the following best reflects your *ability* to cook? (Check one.)

_____ I am a skilled, adventurous cook and can cook just about anything.

_____ I know and can do some of the basics.

_____ I can't boil water.

Which of the following do you *own and use* regularly? (Check all that apply.)

_____ one or more chef's knives

_____ food processor and/or blender

_____ measuring spoons

_____ measuring cups

_____ instant-read thermometer

_____ vegetable peeler

_____ whisk

_____ stockpot

_____ one or more saucepans

_____ one or more skillets

_____ wok

_____ one or more baking pans (including cookie sheets)

_____ colander

_____ hand or countertop mixer

In the context of improving your food knowledge and skills relative to health, list foods or menu items that you would like to learn to prepare.

List flavorings such as herbs and spices that you would like to learn to use.

List cooking skills and techniques that you would like to learn.

List kitchen tools and equipment that you would like to learn to use.

Describe any other kitchen and cooking knowledge that you would like to learn.

TABLE 8. Cooking Class Schedule for Wellness Program, Fall 2014

Food for Health Wellness Program

UW–Madison Police Department, in partnership with the UW–Madison Department of Food Science

(Collaborators: Cherise Caradine, Jeremy Isensee, Mark Silbernagel, and Monica Theis)

Date	Focus	Activity
9/17	Introduction and identification of learning topics Review of lab safety	Tour of lab
9/24	Knife skills: selection, use, and care	Preparation of salads and salsas
10/1	Vegetables: roots, tubers, bulbs, stems, and leaves	Variety of vegetables
10/8	Protein: animal-based	Roasts, steaks, filets Beef, pork, and fish
10/15	Protein: plant-based	Tofu, tempeh, seitan, beans, nuts, and other legumes
10/22	Soups, stews, and sauces	Marinara and pesto "Better for you" chicken noodle soup
10/29	Breakfast slow; breakfast on the go	Slow-cooked whole grains; breakfast bars
11/5	Lunch and dinner on the go	Healthy bag lunches Cooking without a plan Planning and shopping
11/12	"Better for you" favorites	Macaroni-and-cheese appetizers
11/19	Real and naturally sweet desserts	Variety of fruit-based desserts

in learning how to provide healthy food for their families. All participants were interested in learning to prepare meals that were quick, healthy, delicious, and compatible with a stressful work life, odd work hours, and busy lifestyles.

To gain an appreciation for specific learning needs and requests, a survey was distributed to all participants (table 7). The survey feedback was used to develop the pilot class series, which consisted of ten two-hour classes on Wednesday evenings (table 8).

TABLE 9. Food Questions Raised in Wellness Program Classes

Food for Health Wellness Program
UW–Madison Police Department, in partnership with the UW–Madison Department of Food Science

How should I store fresh vegetables? Which vegetables should be stored at room temperature?
Is organic food healthier? What does "certified organic" actually mean?
Isn't our soil depleted from conventional agriculture, so vegetables have fewer nutrients now?
Can we talk about juice and other beverages relative to health? For example, is orange juice as good as eating an orange? How much juice should we drink?
Are there tips for how to pack a healthy brown-bag lunch?
Tell me how to stock my pantry and cupboards so I can make good, healthy meals when I come home, even though I am tired from the day and don't have a meal planned.
Should I buy a food processor? If so, which one?
What should I eat and drink after I work out?
What do I look for when selecting fresh garlic?
How many and what types of cutting boards should I have in my kitchen?

Questions from the Meat Lab
What do I look for when selecting and buying fresh meat?
What is the best thermometer for checking temperatures of food for doneness and safety?
To what temperatures should various meats be cooked for safety? Best yield? Done?
Should fresh pork be frozen for safety reasons when brought home from the store?
If pork is pink after cooking, should I throw it out?
How long can various meats be kept fresh before being cooked or frozen?
How long can various frozen meats be kept?
What are some recommended recipes for ceviche?

The pilot program was exceptionally successful. Each week the class began with ten to fifteen minutes of discussion, followed by an introduction to the cooking focus of the evening. The participants worked in pairs, preparing a recipe and then sharing the product with the instructors and fellow participants. Guidance was provided by two instructors and volunteer students from the dietetics program at UW–Madison. Each session ended with an opportunity for the participants to ask

TABLE 10. Cooking Class Series for Wellness Program, 2015

Food for Health Wellness Program
UW–Madison Police Department, in partnership with the UW–Madison Department of Food Science

	Primary Focus	Secondary Focus	Activities
1	Program introduction Knife skills	Flavorings: vinegars and oils	Salads and salsa
2	Vegetables: roots, tubers, stems, leaves, and seeds	Vegetables from the sea Cooking methods: roasting, sautéing, stir-frying	Preparation of a variety of seasonal vegetables
3	Protein: animal-based Beef, pork, poultry, seafood	Flavor enhancers: rubs Cooking methods: poaching	Preparation of a variety of cuts
4	Protein: plant-based	Turning up the heat: Chilies and Scoville heat units Tree nuts	Preparation of a variety of entrées using meat analogues, such as beans and other legumes
5	Stocks, soups, stews, and sauces	Pasta and pastalike alternatives (e.g., zucchini) Flavoring: aromatics	Preparation of cream and broth-based soups Preparation of the "mother" sauces
6	Breakfast slow; breakfast on the go	Lunch and dinner on the go	Preparation of whole grains, breakfast bars, and breakfast burritos
7	"Better for you" favorites	Flavoring: peppercorns (red, yellow, pink, and green)	Your favorites with a health twist
8	Sweet and natural: cooking with fruit	End-of-the-program celebration	Preparation of fruit-based desserts using berries, stone fruits, and more
	Other: Freezer-friendly foods Kitchen essentials (tools and gadgets) "Better for you" fats The well-stocked pantry (staples for busy people)		

questions and share ideas for program improvements. The instructors did their best to keep track of questions and comments in an effort to adapt the program as it unfolded and to plan future classes. Table 9 shows a list of questions raised by the participants.

Testimonials from the participants attest to the success of the program. In addition, a brief survey distributed three months after the program indicated that the UWPD staff members were continuing the food selection and preparation changes they had made as a result of the first series of classes.

The class series continues to be offered. The participants from 2014 were offered the option to join the class again for additional food and skill instruction. New UWPD staff members were invited to participate in a series of classes uniquely designed to their needs; the 2015 class series is shown in table 10.

CONCLUSION

Many consumers have begun to make food purchases in the spirit of health, environmental stewardship, and support of local economies. Lack of knowledge about food and its preparation presents barriers that can be overcome through thoughtful and dedicated education programs that focus on the unique needs of various consumer groups. Programs such as those developed for MOM and the UWPD provide a basis for identifying the content, materials, and resources that are necessary to effectively influence cooking in the context of lifestyles and food values.

CHAPTER EIGHT

It All Starts with the Soil

STEVE VENTURA

In this chapter Steve Ventura focuses on composting as the most important way of returning food waste to food production—the resource and waste recovery component—to complete the cycle of elements in the food system supply chain and highlight a significant contribution of food system actions toward overall environmental sustainability.

A s Will Allen of Growing Power travels around the country, spreading his ideas and methods for urban food production, he invariably says that "it all starts with the soil," referring to his organization's nutrient-rich compost and vermicompost. Vermicompost results from the additional working of compost by Allen's "livestock," red wiggler worms. To amplify his point, he has created productive gardens directly on pavement by putting down a thick layer of wood chips followed by at least eighteen inches of compost (fig. 13). Although the long-term efficacy of this practice is unknown, and water management is tricky, it makes the point that the medium is the message, and in a similar way that Marshall McLuhan intended his famous saying originally: if we pay attention to the soil (the medium), other components of community food systems can be changed, too (the message).

Building a good growing medium—whether it is carefully managed native soil, compost, vermicompost, or even lightweight engineered soil suitable for rooftops—is a response to three issues in urban agriculture. The first is obvious: plants grow better in a root environment that readily provides nutrients, water, air, and a stable base. The second is an unfortunate but common problem: the native soil of urban areas may be contaminated, compacted, or in other ways de-

FIGURE 13. Growing Power renews an urban garden by layering fresh compost over old at the Badger Rock Center, Madison, Wisconsin. (Courtesy of Martin Bailkey, CRFS)

graded by decades or centuries of human development and abuse. The third is a by-product of another issue: We produce huge amounts of organic waste. By some estimates, as much as 40 percent of food produced in the United States is uneaten (Gunders 2012). Landfills and wastewater treatment plants handle most of this food, but they bury or destroy valuable nutrients and soil-building organic matter in the process as well as contribute to greenhouse gases and other environmental problems (Hall et al. 2009).

Recycling unused food and other organic waste through composting directly addresses all three issues. Because composting methods exist for a wide range of scales, they can be used for a wide range of urban agriculture projects, from backyard gardens to large commercial operations. In community food systems, composting becomes the key link between the last step of the food cycle, which is consumption, and the first step, which is identifying and creating fertile land. Composting brings consumers and growers together again.

As part of the CRFS project, we focused on how both pre- and postconsumer food waste was turned back into growing media as well as on the community food

FIGURE 14. Bin-scale vermicomposting. Households can compost kitchen scraps in five-gallon buckets. With drainage, aeration, and management of feedstock, these buckets can support healthy populations of red wiggler worms to create rich vermicompost. Here Will Allen shows off his bin-scale vermicomposting. (Photograph by Michael Kienitz, courtesy of College of Agricultural and Life Sciences, University of Wisconsin–Madison)

system choices that supported those processes. We investigated food waste conversion in our project cities, and in the course of pursuing these issues we learned about efforts in additional communities. These communities generated lessons about various systems and motivations for converting food waste to soil. When we began this inquiry we expected to obtain clearly articulated principles and distilled indicators of success. What we found instead was a field of experimentation and engagement at all levels and sectors.

Numerous books, bulletins, websites, training sessions, and other resources describe *how* to compost, often providing appropriate options for growers, communities, or waste managers based on scale and circumstance. Instead, this chapter studies how composting fits into community food systems. After a brief review of the biophysical processes and environments, this chapter discusses composting's economics, community and cultural interactions, policy perspectives, and questions about health and nutrition.

FIGURE 15. Pallet composting. Community gardens can build simple structures such as these five-pallet crates to collect scraps and slowly compost them. Because they aren't turned (aerated), decomposition is slow. Multiple bins may be needed to support the sequencing of production. (Courtesy of Greg Lawless, CRFS)

THE COMPOSTING ENVIRONMENT: CAPACITIES AND CONSTRAINTS

A systems perspective on composting includes biological, mechanical, and environmental processes. Successful operations must consider all three, in addition to social aspects such as economics, labor, and policy. Each of these domains comes with multiple options for communities to consider in light of their existing capacities and desired outcomes, along with biophysical and social constraints. As a result, successful composting can occur at many scales through many approaches, but possibilities abound for less than successful outcomes.

Composting is a natural biological process in which microorganisms obtain material and energy from organic matter. In open composting, this process involves the aerobic (with oxygen) decomposition of organic material by a suite of bacteria and fungi. The result should be soil-like material, rich in organic matter, that includes ammonium or nitrates (nitrogen fertilizers necessary for plants) and

FIGURE 16. Wind row composting—farm equipment. The production of large volumes of compost requires mechanical equipment for mixing and turning. At the Farm near Sturtevant, Wisconsin, a dump truck delivers food waste, and a bucket loader or skid steer is used to mix and turn compost piles. (Courtesy of Susan Bence, WUWM Milwaukee Public Radio, and the University of Wisconsin system board of regents)

other plant nutrients, depending on the feedstock. It should be free of significant amounts of contaminants or detritus.

Feedstock

Under the right conditions, microbes can break down almost any kind of feedstock, including food scraps, food processing by-products, and other organic materials such as shredded leaves or wood. The composition of the feedstock, primarily the amount of available carbon relative to the amount of nitrogen-containing compounds (e.g., proteins), influences the rate and products of breakdown. Environmental conditions, particularly moisture and air levels, are also an influence.

Many variables affect the quality of compost: feedstock, material mixing and handling, aeration, the ratio of carbon to nitrogen, moisture levels, and time are the most important. To create a safe, consistent product requires consistent and regular management. This is a key challenge for composting operations. Methods

FIGURE 17. Wind row composting—special equipment. Large volumes of compost are managed in long "wind rows." Heavy equipment builds and turns piles at a large-scale operation at the University of Wisconsin West Madison Agricultural Research Station. (Courtesy of Steve Ventura, CRFS)

for managing these variables exist across a wide range of scales, from household composting in five-gallon buckets (fig. 14) or wood bins to midsize pallet crates typical of community gardens (fig. 15) to massive private and municipal operations that handle tons of material daily (figs. 16 and 17).

Getting the right feedstock for composting is both a logistic (material handling) and biological challenge. Although it may be desirable in theory to return all food waste to growing systems, this is not practical. For example, postconsumer food waste (after the food has been on a consumer's plate) is considered hard to handle, messy, and a potential health hazard. In some communities it is illegal to include postconsumer food in compost. This stigma is a result of the highly variable composition, including the likelihood of odor-producing material such as meat scraps, nondecomposable material such as plastic, and infectious disease organisms.

The easiest way to control feedstock variability is to use only pre- or nonconsumer materials. In urban settings this includes yard waste, coffee grounds, spent grains from beer brewing, preconsumer trimmings from restaurants and institutional kitchens, and waste from supermarkets and wholesalers, such as fruits and vegetables that have ripened or spoiled beyond the point of salability. In the context of an urban food system in which food access and affordability are issues, the latter source should be carefully considered. Options for bringing salvageable portions to shelters, pantries, and other distributors of emergency food should be part of the collection process.

Careful use of pre- and nonconsumer materials can lead to high-quality compost if handled well. For example, Purple Cow Organics is a soil and landscape business in Madison. The company collects food residuals and by-products that are "clean" (e.g., no meat, bones, or animal fats that can become rancid and odorous) and generates premium compost sold to certified organic growers.

Source separation is an important approach to the variable composition challenge of postconsumer waste. However, it requires an educated consumer who doesn't just dump meal refuse into whatever container is handy. Multibin collection is necessary, and it typically requires at least three containers: compostable, recyclable, and waste (garbage taken to a landfill). In many situations it cannot be assumed that consumers will learn the subtleties of which food waste can or cannot be composted; simply collecting all food waste and compostable paper may be the only reasonable goal. In this case a composting system that can handle high-protein waste is necessary.

Three approaches are used to deal with the challenge of mixed waste: postcollection separation, composting with odor and pest control, and anaerobic (lacking oxygen) digestion. Obviously, postcollection separation is a messy, smelly process and generally not suitable for large volumes of material. Unseparated material can be composted if it does not contain large amounts of nondecomposable material such as plastic and treated paper. It requires careful attention to compost conditions—aeration, moisture, and the ratio of carbon to nitrogen—to make sure the material composts completely and at a sufficiently high temperature to kill pathogenic organisms. Insect and rodent control may also be necessary.

So-called compostable plates, cups, and silverware can present a challenge for composting postconsumer material at any scale. Although paper eventually breaks down, the process is slowed by biodegradable waxes and coatings on cups and plates. These items should be avoided in any significant amount in systems that do not achieve consistent high temperatures and rapid breakdown conditions. These items should also be avoided in home and community operations. Several companies market "compostable" silverware made primarily from biodegradable corn or potato starch, but numerous investigators have found that the silverware does not break down at the same rate as most other compostable material. Large-scale operations may need a postcompost "finishing" phase of up to 180 days to sufficiently break down these bioplastics. A bulletin by the State of Washington's Department of Ecology provides a useful summary of the potential and limitations of biodegradable and compostable serviceware (Xiao 2014).

A few communities and organizations have successfully incorporated anaerobic decomposition into their operations. This requires specialized equipment: essentially large sealed drums that contain material and gases while the material is broken down by anaerobic organisms. The resulting products include methane gas, which can be captured and used for energy, and decomposed organic matter that can be used directly as a soil amendment and fertilizer or added to compost. Anaerobic digesters are very useful for handling high-protein material that would ordinarily be avoided in regular composting, but they are comparatively expensive and require a high level of management to maintain and operate the machinery.

Home-Scale Composting

STEVE VENTURA

Not everyone has the luxury of land or a convenient community compost bin. Households may not generate enough organic waste to attain critical mass. Without a few cubic yards of well-aerated material, the compost process won't generate enough heat to kill weed seeds, and the process can take several months. Worm buckets and other forms of vermicomposting can accelerate processing but shouldn't generate a lot of heat.

Another option for households is a commercial composting machine, such as the NatureMill indoor automatic composter or the Ecotonix GreenCycler. As with any composting, these machines require some knowledge of the process to avoid a fetid, slimy mess.

Sending food waste down the drain to a wastewater treatment plant via a food grinder (e.g., InSinkerator) has been suggested as an eco-friendly alternative to composting. This approach may be better than sending food waste to a landfill, but it is eco-friendly only when the treatment plant is meticulous about collecting the methane generated by its treatment processes, when the resulting biosolids are returned to the land, *and* when the biosolids are not contaminated by all the other things that are flushed down drains. Food grinders have been used creatively, however. In backyards with limited space for composting, do-it-yourself enthusiasts are experimenting with food grinders mounted over buckets or tanks for home-scale composting. The finely ground and moisturized material will break down very quickly if an appropriate carbon-to-nitrogen ratio is maintained. For an example, see Natchez (2011).

Feedstock Contaminants

The potential for contaminants in feedstock must also be considered on both a chemical and a biological level. "When composting on a smaller scale, pathogen barriers are of more concern, because the smaller the pile gets, the more challenging temperatures are to control," according to Joe Van Rossum, recycling specialist and director at University of Wisconsin Extension in Madison. Compost has to get hot enough to kill undesirable microorganisms and seeds.

Residual herbicides can be an important yet easily overlooked barrier to community composting. Herbicides are used on crops and may find their way into composting as sources of both carbon (from straw or dried leaves) and nitrogen (from animal manure). Lawn and yard waste can also contain residual herbicides from lawn-care applications. Most herbicides break down quickly in the environment. However, the National Center for Appropriate Technology warns about a couple of particularly persistent herbicides: Clopyralid, used in turf management, and Picloram, used for broadleaf weed control. These herbicides have a residual life of up to three years after application. The high temperatures reached when composting are not enough to affect residual herbicides; in fact, the composting process concentrates it.

Herbicides leaching from composting areas within community garden sites find their way into individual plots and can damage or kill plants. More awareness is needed of the provenance and quality of source materials for compost whenever

FIGURE 18. The Full Cycle Freight bicycle-based compost operation created by the F. H. King Students for Sustainable Agriculture, University of Wisconsin–Madison. (Courtesy of Bradley Meilinger)

the goal is for use in urban agriculture, community gardens, or other agricultural settings. The US Composting Council is trying to raise awareness of this issue and alter the EPA registration requirements for the most persistent herbicides.

Collection and Compost Operation Siting

Compost, food waste, and other feedstocks are large-volume, low-value commodities. Efficient collection and distribution systems can be critical to their success. The compost needs of a home or small community garden may be met by a convenient collection site in close proximity to the growing site. For example, Kompost Kids helps maintain seventeen feedstock drop-off sites in the Milwaukee area, most consisting of simple wood-frame bins. The F. H. King Students for Sustainable Agriculture at UW–Madison collect compost from the campus area in bicycle trailers and use it at their student farm (fig. 18).

At a midrange scale, Growing Power in Milwaukee has made the collection of pre- and nonconsumer waste its mainstay for creating compost. Every day trucks pick up and deliver material to a central compost yard. This process requires a truck dedicated to hauling raw material and finished products, a front-end loader to move and turn the compost, and one or more Growing Power staff members dedicated to the compost operation. Will Allen attributes the success of this approach to the reliability of pickup, even when conditions are nasty.

Composting for an entire municipality requires a collection system that can move large volumes of material cheaply. The easiest alternatives are the business and institutional sources that ordinarily pay directly for waste hauling. Because compost has some value as a product and reduces the filling of expensive landfills, large operations can provide a less expensive alternative to landfill tipping fees (charges for waste disposal) if the businesses or institutions can provide separated organic matter. For example, in 2014 the Solid Waste Agency of Cedar Rapids/ Linn County, Iowa, charged a $38 per ton landfill tipping fee and $18 per ton to process food scraps and yard waste. This has actually turned into a public service; area residents can receive up to a ton of well-finished compost daily by having it dumped by agency loaders into their pickup trucks.

Residential collection of organic matter is a significant challenge. Separate collection bins are a must, but it is difficult for consumers to remember what is trash, what is recyclable, and what is compostable. San Francisco has one of the best examples of a three-bin system; the city has included incentives in the collection system fee structure to encourage source separation and trash reduction. A report by the City of Milwaukee (Meyers 2015) found that a three-bin system would add $7.70 a month per household to their collection costs. This was not enough to break even, if only landfill tipping fee reductions were considered; the report did not account for the value of the resulting compost or other potential benefits.

In interviews for the CRFS project's evaluation of food waste, conducted by interns with the Michael Fields Agricultural Institute, composting experts in several communities noted the potential for competition between waste haulers–landfill operators and community compost organizations. Waste haulers have contracts and infrastructure devoted to current waste management systems and may be reluctant to give up a portion of their collection business. A compost collection manager in Milwaukee described the challenge this presents to community composting. Large-scale waste haulers are not interested in separating organic matter because it contributes substantial weight (hauling contracts are weight-based) and generates methane for energy at landfills.

Particularly for large compost operations, the siting of a facility can be critical. Even backyard compost heaps can draw complaints from neighbors about odor, noise, and debris. Growing Power's lease on a site in Oak Creek, Wisconsin, was not renewed because of complaints from neighbors, even though it was a quarter-mile from a wastewater treatment facility and produced only good, earthy decomposition smells, according to witnesses.

In addition to being acceptable to neighbors, a site must have good access and drainage. If the base of a compost pile remains saturated with water for more than a few days, it will become anaerobic and produce objectionable odors. Conversely, a water source may be necessary to keep compost working when rainfall is insufficient to maintain the necessary moisture level.

For operations at the neighborhood and community garden level, Growing Power promotes simple, inexpensive compost bins built from five wood pallets. These square boxes are approximately four feet long per side. The boxes are lined with wire mesh to keep out rodents. Carbon-rich and nitrogen-rich materials are layered into the box until filled. The pallet boxes can be set up where appropriate, and slats allow them to drain easily. The main disadvantage is access—the boxes are not easily mixed or turned for aeration, so the composting process takes a long time. Recently Growing Power has been recommending paired pallet bins. One is filled and then periodically dumped into the other empty bin to overturn the material for aeration and moisture control.

Water flow must also be considered. Dry organic matter decomposes very slowly, so a source of water is a necessity if compost production cannot wait for natural precipitation. At the other end of the spectrum, if rainfall or snowmelt exceeds what a compost pile can absorb, the water will leach through the heap and become rich in phosphorus and nitrogen. In the wrong places, like groundwater and nearby streams or lakes, this nutrient-rich water is a pollutant, so proper drainage and runoff management must be considered.

Ideally compost drainage is captured in a cistern or other collection system and used as fertilizer or later returned to the heap. An adequate solution entails directing the runoff to basins and other open areas with plants that can soak up the nutrients. Another solution is to cover the compost. Large operations can be done in sheds; for smaller operations, tarps and other temporary structures work. Groundwater contamination can be minimized if composting is done on an impervious surface such as a concrete pad or compacted clay, but this approach can add considerable expense to the setup.

To be sustainable long-term, composting must make sense economically as well as environmentally, as discussed in the previous section. However, economic sustainability does not mean that the operation has to pay for itself. Because community and regional-scale composting are done for multiple goals, nonmonetary benefits accrue, including diversion of food waste, increased soil health, participation in community-building activities, and job creation.

Some benefits of composting, such as fewer costs through avoided hauling or landfill tipping fees, are readily monetized. With attention to creating a uniform safe product as well as to marketing and distributing, compost can even have direct commercial value, even though this is not typically the primary goal of small-scale systems. Other goals, such as community benefits, are harder to value, but economists use cost-benefit analysis techniques that can be applied, if necessary, for project evaluation. Buckley and Peterson (n.d.) describe some of the cost-benefit techniques applicable to community food systems.

Of course, operating a system of any size requires real resources for expenses: funds for people, site preparation, pile management, material collection and distribution, equipment acquisition and maintenance, monitoring, regulatory compliance, and so forth. Scaling up generally includes replacing human energy with mechanical energy. Equipment can be a large and even prohibitive cost in expansion. In addition, administrative costs such as record keeping, marketing, public relations, and site security increase with operation size and may affect a community's incentive to compost.

Resources to fund a system typically include a combination of grants, public subsidies, sales revenues, voluntary labor, donated space, and partnerships with waste producers and waste haulers. The mixture of these sources will depend on the type and scale of the facility. Continued success in securing grants and maintaining partnerships will depend on evidence of the operation's effectiveness, feasibility, and sustainability. For many potential supporters, the best evidence is economic (e.g., avoided costs, sales revenue, and job creation), but this should not preclude evaluating and documenting nonmonetary benefits.

The Institute for Local Self-Reliance completed a careful and comprehensive evaluation of the economics of composting (Platt, Bell, and Harsh 2013). Although this study evaluated composting across the entire state of Maryland, many of the conclusions can apply to medium-size and large cities as well. The report evaluated the cost of creating collection programs and maintaining compost facilities. It also provided clear information about the benefits of composting, including reducing wastes, improving soil, creating jobs, and supporting local economies. The comparison of composting versus landfilling or incinerating was quite remarkable: "On a dollar-per-capital-investment basis, composting operations sustain three times more jobs than landfills and 17 times more jobs than incineration facilities in Maryland" (ii). The report also notes that composting works in a wide range of scales and sizes, and, in fact, a mixture of small and medium-size facilities provides diversity adapted to local needs and a superior benefit stream.

This is not to suggest that no large facilities should be part of the mix. As a society we generate huge volumes of organic waste, and cities shoulder the burden of managing it. George Dreckmann, the former City of Madison recycling coordinator, had this management responsibility for a city of almost 250,000 people. "Compost generation falls into a process of economy of scale," he said. "The more material that is collected and processed, the higher the level of efficiency." For example, in Cedar Rapids, Iowa, initial efforts to create a large-volume facility floundered after inception because the operation focused exclusively on waste from supermarkets. The operation then expanded its partnership base by engaging universities and large organic waste–producing businesses, and now the system is sustainable.

COMMUNITY AND CULTURAL RELATIONS

A mixture of small, medium, and large efforts can be supported by conducive public policies and partnerships. Some of our CRFS collaborators reported that a key element of their success was the connection with partners. Others perceived their failure to engage partners in the beginning of the project as a reason they struggled to maintain operations. Partners who were identified as important in food waste management include the following:

- Food waste producers: homes, farms, community gardens, and businesses, particularly large grocery stores and restaurants
- Haulers and facility operators
- Public officials: people developing policies that affect composting
- Research facilities: generators and disseminators of evidence-based knowledge
- Community members: grassroots advocates and consumers

No partner should be left behind, because the cycle of converting food scraps to soil amendments will probably involve every sector's participation sooner or later.

Small-scale efforts such as neighborhood collection sites and community garden composting are important for raising awareness of organic waste management issues and getting people involved. Larger efforts, such as municipal systems, can be important sources of materials for supporting urban gardening and indicators of public agency support for enlightened management of food waste. If the goals include collection of residential waste (at either scale), public information in the form of signage, brochures, and websites can be helpful in generating public participation and support.

Policies that support composting or digestion are critical for these initiatives. Composting may be regulated at local or state levels or both. For example, the Wisconsin Department of Natural Resources has rules for facilities that contain more than fifty cubic yards of material on site at a time. Sites with more than twenty thousand cubic yards must have annual inspections and an approved operating plan. Smaller piles, which are typical of community or household efforts, fall under local ordinances, if they exist.

In general, policies are aimed at preventing nuisance conditions such as odors, pests, and nutrient-laden runoff. The Wisconsin Administrative Code NR 502.12(4)(b) requires that "the facility is operated in a nuisance-free and environmentally sound manner." Awareness of the benefits of composting and the need for compost have created interest in changing overrestrictive rules or limits to what can be done in cities. For example, the Chicago Food Policy Action Council and others succeeded in changing antiquated rules developed primarily for waste management in rural animal agriculture. The council's efforts resulted in a modification of Illinois's Environmental Protection Act, clarifying that composting operations are not pollution control facilities and thus are not subject to the same standards as everyday waste facilities (Suerth and Morales 2014).

It is also possible to create local policies that actively support composting. Suerth and Morales (2014) describe approaches to local zoning that are supportive of composting, either as accessory uses in specific districts or as a component of permitted land uses such as community gardens. Other policies might include subsidies, infrastructure development, purchase or lease of land for facilities, education and training programs, and facilitation and promotion of programs. Local governments should work with community organizations to promote effective solutions to waste management. Local efforts can also work in collaboration with food rescuers, people and organizations that recover usable food for pantries, shelters, and other emergency food operations. (Legal issues related to food safety make this a complicated issue.)

Even a few states are taking on the issue; Vermont, Connecticut, and Massachusetts require large-volume commercial food waste generators to compost or digest their discarded matter instead of dumping it in a landfill. Several other states are considering similar steps. With encouragement from nonprofit organizations and public agencies, the private sector—including haulers, recyclers, and landscaping companies—can become important contributors to this piece of the good food revolution.

Soil scientists have long said, "Stop treating our soil like dirt." Let's add, "Start treating our dirt like soil." Food waste can and should be a valuable addition to community and regional food systems.

CHAPTER NINE

Uprooting Racism, Planting Justice in Detroit

JEFFREY LEWIS, NICODEMUS FORD,
AND SAMUEL PRATSCH

In this chapter, three extension service evaluation specialists tell the story of Detroit residents using food justice as part of an ongoing grassroots effort to dismantle systemic and institutional racism. It is oriented to the CRFS values of justice and fairness.

U prooting Racism, Planting Justice (URPJ) is an organization committed to achieving racial equity and dismantling racism in the Detroit food system by creating a variety of community spaces where people can develop relationships, learn about the structure and impact of racism, heal from the harm that racism causes, and organize. URPJ uses food as one concrete set of issues through which to explore the broader array of racial equity. Its work puts food-related issues into a larger context and provides entry points for people from diverse socioeconomic and cultural backgrounds and perspectives to develop leadership, build capacity, and create accountability for achieving racial equity and dismantling systemic racism in the food system and broader society. URPJ uses training and workshops to educate individuals and groups about the structure and impact of racism in their lives, in food systems, and throughout the Detroit community. URPJ participants create spaces in their caucuses and study groups to develop relationships, heal, and support and challenge one another. They strive for justice through projects and outreach efforts and in their everyday lives.

This chapter highlights the work of URPJ and the outcomes and lessons learned from a project evaluation conducted by the CRFS project team members in 2014.

The collaboration between URPJ and the CRFS project was formally structured as a community engagement project and was designed to identify and execute efforts in which the university partnership could best support grassroots work in the CRFS project cities. In Detroit the community engagement project with URPJ aimed to build a partnership with equal power sharing, with benefits realized both by URPJ and the CRFS project, and that met community needs in effective and practical ways. This type of collaboration and participatory approach was especially important, given the historical and social implications in a city such as Detroit.

DETROIT'S STRUGGLE FOR JUSTICE

Detroit lays bare the limits and contradictions of industrial economies unlike any other city in the United States. Once the fourth-largest US city, with nearly two million people in 1950, it flourished when the automobile industry expanded into the everyday lives of most Americans. It now reflects and embodies a postindustrial reality: the loss of blue-collar manufacturing jobs, a sharp decline in its population and tax base, the demographic shift from more than 80 percent white residents to more than 80 percent black residents today, and the nearly catastrophic withdrawal of public funding for infrastructure, schools, health, and other social services necessary to ensure the basic human rights of its citizens.

In the first half of the twentieth century, Detroit was one of several midwestern cities where industrial labor forces were built in large part on the Great Migration of African Americans from the US rural South. Blacks sought a better life and created communities in their new urban neighborhoods. Thomas (1992) defined the process of building a community among African Americans in Detroit as the total efforts of black individuals and organizations to survive as a people and to create and sustain a genuine communal presence. Despite the loss of more than half of the city's population in recent decades, many Detroit residents have responded to these challenges with resilience and imagination.

Detroit has a long, rich history of black and working-class struggles for racial and economic justice. Whether it was black laborers organizing themselves for more just and fair working conditions in Henry Ford's automobile factories, or the Resistance Movement of 1967, when Detroit residents rebelled against a racist police force, the city has long-established roots of resisting oppression. According to Sugrue (2005, 89–90), African Americans "struggled for rights, justice, and power on a terrain that was shaped by the vision for urban change that ultimately influenced public policies from affirmative action to anti-poverty efforts to education reform, even if they struggled against the odds in a political climate that grew increasingly hostile to their demands."

This history helped establish the social relationships, structures, processes, and activism that fostered and nurtured thriving and resilient black communities in Detroit. The activist food security and food justice work of URPJ extends this history into the present situation in Detroit, through the commitment by its participants to develop and sustain "a genuine and creative communal presence" to improve the lives of black and working-class people.

Detroit, like many other American cities, has found itself "burdened with an aging infrastructure, an increasingly impoverished population, and fewer resources than ever to pay for infrastructure repairs, education, or social services." This has led to "a reallocation of political power and public resources to the increasingly privatized, exclusionary world of white suburbia" (Sugrue 2005, xxii). The flat and declining incomes of previously working-class and middle-class families, and the erosion of neighborhood and community institutions, have left many Detroit residents "mired in poverty-induced challenges" that include "little or no access to healthy food" (White 2011, 14).

Citing Zenk et al. (2006) and Gallagher (2007), White pointed out that, on average, residents of African American communities must travel more than one mile farther to get to a supermarket than people living in predominantly white communities. However, many low-income individuals and families in Detroit lack personal automobiles, and continued service reductions by the city have made public transportation less reliable for many. Limited incomes and the lack of reliable and affordable transportation, combined with long distances to supermarkets, result in an overreliance on fast food and convenience stores in low-income black communities. More than 80 percent of Detroit's residents depend on these stores (Gallagher 2007), which is directly linked to a decline in consumption of fruits and vegetables by African Americans (White 2011) and ultimately results in poor health outcomes and poor school performance.

STRIVING FOR FOOD JUSTICE THROUGH ANTIRACISM WORK

URPJ operates according to the mantra of Malik Yakini, the executive director of the Detroit Black Community Food Security Network, that "There can be no food justice without social justice" (Civil Eats 2016). URPJ directly addresses food insecurity and food justice by embedding its antiracist work within the context of Detroit's current food system. However, URPJ's work goes beyond the struggle for food security and reflects the characteristics of food sovereignty (see chapter 1). Proponents of food sovereignty assert that food is a basic human right and that, as stated by La Via Campesina, everyone has the right "to healthy and culturally appropriate food produced through ecologically sound and sustainable methods, and [a] right to define their own food and agricultural systems" (Alkon and Mares 2012). The notion of food sovereignty acknowledges the systemic and structural inequities within commodities markets and food systems. In addition to centering access to food and control of food choices as a basic human right, food sovereignty affirms the importance of the health of cultures and places and the right to self-determination.

URPJ's mission, values, and approach, and the activism of many of its participants, reflect a combination of food access, food security, and food sovereignty. URPJ centers its racial justice work on the following values:

1. A food system that reflects the population, its expertise, and its values
2. Historical traditions and cultures
3. The power of the community within the food system
4. Accountability at all levels: individuals, institutions, and systems
5. Healing as an integral part of antiracism and antioppression work
6. Just and equitable policies that inform the practice, governance, economics, and allocation of resources throughout the food system

URPJ plays an important role in helping Detroit's black community and food system activists to develop a vision for just and sustainable ways of living. For many of its participants, URPJ continues and builds on this rich history of struggle and leadership.

THEORY OF CHANGE

URPJ formed in 2010 to foster commitment, education, and activism that would achieve racial equity by dismantling racism at all levels of the local and regional food system. URPJ members commit themselves to "advocacy for accountability and just policies; training for leadership development; and building power through collaboration, organizing, and mobilization."

URPJ developed a framework it found useful to address racism in the food system and bring about broader social change. The framework portrays how individual change combined with community connections, healing, and shared struggle serve as the foundation for antiracism activism and systems change. Four central elements of URPJ's theory of change build the individual and collective capacity to achieve these outcomes:

1. Acquiring a critical knowledge of individual and systemic racism
2. Having meaningful conversations about race and developing conscious relationships
3. Developing shared language and frameworks
4. Critical reflection and organizational analysis

Collectively these four factors foster awareness and a shared analysis and understanding of racism in the food system and the society at large. Acting on the factors can lead to changes in behavior and in antiracism advocacy and to activism by individuals within their organizations, which in turn can lead to equity and justice. URPJ hopes that its theory of change, through antiracism training, monthly gatherings for its members, and supporting caucuses (study groups), leads to initiatives to challenge and dismantle racism in Detroit's food system.

TRAINING IN DISMANTLING RACISM

Working primarily with the curriculum and staff of the People's Institute for Survival and Beyond (PISB), an international collective of organizers and educators, URPJ conducts training to help people in Detroit understand how racism affects or is part of all social, economic, and political systems. These training sessions provide a conceptual framework, analytical tools, and group experiences. The participants examine the systemic nature of racism, with a focus on how various forms of oppression combine and intersect in everyday life (intersectionality). They discuss power and organizational gatekeeping within systems and learn more about "the historical community" to better understand the circumstances, experiences, and struggles with which people live and to help guide action.

At the heart of PISB's antiracism training is the analysis of how power is embedded in structures and systems of oppression that maintain white wealth, power, and privilege. The power analysis helps the participants understand the historical dimensions of power and inequality and their contemporary expressions at all levels of society. It also provides the participants with a tool to critically analyze the systemic, historical nature of racial oppression and how it is reflected in contemporary social and economic inequality. In addition, the power analysis helps the participants understand the "cumulative effects of institutional, cultural, linguistic, military, and normative oppression of communities of color over time [and] the widespread disempowerment, disorganization, and perpetuation of poverty within these communities" (Shapiro 2002, 10).

PISB training emphasizes leadership, the sharing of knowledge, accountability to communities, the creation of effective networks, and the role of organizing. It also serves as the impetus for developing a social infrastructure made up of monthly Saturday gatherings, caucuses, and direct-action initiatives. Through these structures and within these relationships, URPJ members apply lessons of the training to social change efforts, promoting and building alternative, nonhierarchical, just, and sustainable systems.

SATURDAY GATHERINGS

One Saturday each month, URPJ participants and other Detroiters gather in a circle to build relationships and counter separateness and isolation. These gatherings create a space in which the participants can openly examine the influence of racism in their lives, which the project director, Lila Cabbil, describes as "reconnecting us to our humanity." In addition, the gatherings help the participants learn about and discuss a variety of political issues related to the manifestation of racism in Detroit, particularly those related to ongoing attacks on basic human rights, the growth of gentrification and continued displacement of low-income people, political disenfranchisement, and the public-sector austerity that has a disproportionate and negative effect on black and low-income residents. The participants recognize that these issues intersect with the right to healthy and fair food in many ways.

For example, the participants at one gathering discussed how food is tied to the concentration of wealth and how the racialization of wealth affects inequality and food justice. They then discussed what they could do about the problem as individuals and examined the ways that participation in the economy supports oppressive corporate institutions. The participants then examined how concentrated wealth manifests itself in Detroit (i.e., public resources moved to private hands), which led to suggestions for action and the identification of local groups to connect with to combat the trends toward concentrated wealth and the privatization of public resources.

CAUCUSES

URPJ created and supports three caucuses, or study and discussion groups, for people who attend the Saturday gatherings: the White Caucus, the People of Color Caucus, and the Black Caucus, which are formed around the shared identities, social status, and histories of their members. Each caucus evolved independently of the others and, as a result, has developed a unique identity, agenda, and approach to fulfilling URPJ's purpose and goals. Consistent with the participatory and egalitarian principles of URPJ, each caucus forms circles and incorporates deep listening, dialogue and discussion, and critical reflections. Each also grapples with the manifestations and implications of power in related but different ways.

The White Caucus: Allies in the Struggle for Racial Justice

The White Caucus meets monthly to discuss and work through issues that the participants bring to the group. They also spend time examining the extent to which they feel the effects of power relations on their food justice work. A key focus is to reflect on and find ways to be effective allies to people of color without paternalism, appropriation, or the reproduction of power inequities. Caucus members focus outwardly by discussing how to effect change in their workplaces, boards, and neighborhoods and how to raise the general consciousness of other white people. They focus inwardly through critical reflection and questioning of their own behavior—both within the caucus and as a regular practice in their private lives.

The members of the White Caucus look for ways to apply their experiences from URPJ's antiracism training. Collectively they strive to "communicate these issues with other white folks." They seek tools and ideas that will help them broach the issue with others, continually undo white privilege, and bring the lessons of the training to other groups. They repeatedly express and demonstrate their commitment to staying in relationships with people of color to remain accountable to the project of dismantling racism in the food system and in their communities.

The White Caucus remains challenged, however, by the fact that its relationships with people of color are somewhat tenuous and limited. The members struggle with the question of whether or how they were complicit in the city's gentrification and the resulting displacement of black people, especially if the members are relative newcomers to Detroit. Many members work on this problem by living and working among people of color (usually black people), but that does not relieve them of the privilege they enjoy as white people. As one participant put it, "I just have to be about this work because there are people dying . . . and food is something you can get your hands around."

The People of Color Caucus: Healing from Historical Trauma

The People of Color (POC) Caucus, composed mostly of nonblack minorities such as Latinos and Native Americans, focuses on individual and collective healing from the historical trauma that racism inflicts on individuals and communities. The members apply this work to address food justice while also developing alternative entrepreneurial and economic activities rooted in the group's values of collective wellness and self-sufficiency for a viable local economy.

The POC Caucus meetings provide a space in which individuals can identify how racism affects their lives and those of their families and communities. Two questions guide their work: "How do we create nonhierarchical ways of working together to develop a vision of change and accomplish shared goals?" and "How can we be intentional about healing from historical trauma?"

The POC Caucus meets the third Sunday of each month as a fellowship grounded in "the tradition of using food to connect people" through a focus on shared histories of being both victims and perpetrators of racism. As people of color (most of them the children of immigrant parents), they live with a tension between having positions of relative power and privilege while also experiencing subordinate racial status in the United States, particularly in Detroit, where segregation and city boundaries have locked ethnicity into particular neighborhoods.

The POC Caucus also sees its work as an attempt to stabilize the community through fellowship and grounding. In addition to its gatherings, which center on healing through collective processes, including remembrance, forgiveness, and grace, the members discuss ideas for creating egalitarian, socially and environmentally sustainable ways of living and working together through participation in the new economy movement, which includes cooperatives and small-scale enterprises that meet local needs. In short, their response to racism and oppressive power is to create healing spaces and alternative systems outside or on the margins of dominant social and economic systems in order to redefine and foster collective empowerment. They are challenged to some extent by the limited ability to meet all the needs of people through alternative and informal economies, but a commitment to sharing and simplicity helps. Centering relationships and healing also appear to mitigate this problem. In this way, healing is a foundational principle to undoing racism in the food system.

The Black Caucus: Self-Determination and Economic Development

The Black Caucus meets monthly. Black economic development is the principal focus of the group, with a stated goal of "exploring cooperative economics and addressing land grabs and Detroit wealth extraction." Within this context the members discuss questions about economic self-determination in black communities and black entrepreneurship and debate the role of cooperative economics and cooperative enterprises as a strategy to address the historical and structural nature of racial violence and injustice in the food system and broader economic system.

The Black Caucus focuses on three issues: healing, economic development, and building stronger relationships with immigrants and other disenfranchised communities. The caucus also addresses historical trauma, with a focus on the ways that black people have been historically excluded from and exploited by mainstream social and economic institutions. Caucus members advocate for spending black incomes with black businesses to keep dollars in the hands of black people and circulating in black neighborhoods. The number of black-owned enterprises identified by the group reflects a significant diversity of services and goods, and some geographic diversity in the city as well. D-Town Farm, an independent project of the Detroit Black Community Food Security Network, is leading an effort to organize a food cooperative and business incubator, a planning effort funded as a CRFS Innovation Fund project. This initiative would serve to develop community-owned enterprises that circulate wealth primarily within the community, promoting black "common wealth" and self-determination. In this way, the network's response to historical and structural racism is to unite black people to create black infrastructure and meet the material needs of black communities. However, some caucus members express concern that black business owners operate with the same business logic of the dominant capitalist society—that is, by extracting wealth rather than building shared wealth. Thus there is no assurance of benefits to black neighborhoods and communities.

The discussions among caucus members echo debates within previous social movements and organizing efforts aimed at empowering black people. Their approach reflects the historical tension between the conservative black economics of Booker T. Washington and the Black Panther Party's more radical community-based and community-owned socialist approach as an alternative to the dominant capitalist structure. The specter of internalized racism and the tension of internalized capitalism (a more limited view of "alternatives") versus a desire for radical change are regular points of discussion.

The members of the Black Caucus express passionate, creative, and improvisational approaches to the restoration of economic and community life in Detroit's black communities. The Black Caucus is a testament to and an expression of a fundamental faith in black people and communities that affirms and finds delight in black cultural history and black humanity, whatever the challenges and contradictions. These approaches, conversations, and reflections are only the beginning of examining and undoing racist approaches in food systems.

INITIATIVES AND SOLUTION-ORIENTED ACTIVISM

In addition to training, gatherings, and caucuses, a number of URPJ initiatives are aimed at directly addressing racism, economic self-determination, and food security. One example was Detroit Wheels for Food Justice (DWFJ), a pilot project designed and implemented in 2014 to provide transportation for senior citizens to shop at Detroit's last black-owned grocery store, whose viability was threatened by big-box stores and other corporate chain food markets. URPJ partnered with local churches to use church buses that sat idle during the week to provide transportation for seniors affected by severe reductions in public transportation.

According to Clynes (2013), "One-quarter of Detroit seniors lack access to an automobile; the public bus system and its schedule are challenging and unreliable. Yet seniors need reliable and accessible transportation to get to doctor appointments, to stay connected to friends and families, and to access needed goods and services." Similarly, the Fair Food Network (2013) found that 22 percent of 224 Detroit respondents identified the lack of transportation as the biggest issue they faced in feeding healthy food to their families.

DWFJ was developed, directed, and implemented by "Margaret" (a pseudonym used to protect her identity), an African American woman with substantial social justice experience in African American communities. At the time, she was involved in organizing economic and community development in Detroit, and she believed that the project allowed her to apply skills and experiences developed in her prior activism to the issues of food access and food justice. In her view DWFJ aimed to reorient the expectations of Detroit's senior citizens away from the big-box stores to "a commitment to salvaging" Detroit's last black-owned store. DWFJ organized volunteers and used funds to secure transportation through the church buses. Margaret noted a number of benefits of the project. First, DWFJ connected seniors and low-income families to affordable, good-quality food. Second, the project had an impact on DWFJ's church partners: they became "solution-focused" rather than focused on what is wrong or not working. Finally, the project helped the churches see how to use existing resources in new ways to address a community need.

DWFJ not only helped seniors overcome the access barrier resulting from inadequate or unaffordable transportation, it also eased social isolation, got seniors into other parts of the city, and allowed them to help Detroit's only black-owned grocery store remain in business. Margaret described DWFJ as "reciprocal—a lifestyle of Kwanzaa and cooperative economics," an expression of the politics of black self-determination and an engagement and support site for white allies, partly fulfilling the need expressed by the White Caucus. The project provided white allies with concrete and specific ways to actively support black activism while addressing one aspect of systemic racism in Detroit's food system.

Applying the Curriculum in the Context of Participants' Lives

The participants in URPJ's antiracism training, gatherings, and caucuses possessed a wide range of backgrounds and experience with racism in US society before their URPJ involvement. They also represent a variety of backgrounds in direct antiracism activism and the struggle for racial justice. Some have little or no prior background in antiracism work, whereas others have been involved for decades. Some with extensive background in the struggle for racial and economic justice have not previously applied their backgrounds to food security and food justice activism.

To better understand how URPJ participants apply antiracism training in their lives and work, the CRFS project conducted in-depth interviews with six individuals from the three caucuses. Despite their differences, each participant identified specific ways that URPJ had contributed either to their understanding of racism and white supremacy or to their ability to apply key concepts to specific situations within the local food system or their lives. The participants also connected their work in food security, food justice, or food sovereignty to efforts for broader social change, ranging from neighborhood revitalization and community building to social policy.

Analysis of the interview data suggested that URPJ's program structure and curriculum provided benefits to the participants, regardless of their experience with or knowledge of racism. Three interviewees reported that URPJ helped them develop a specific, solution-based analysis and understanding of what racism specifically looks like where they live and work and that it improved their communication with others.

One of the interviewees, an African American man with decades of experience addressing racial injustice and other forms of social injustice, found his work with URPJ highly beneficial. He noted that URPJ, and specifically the training to unlearn racism, "provide a power analysis to understand the contemporary manifestations of white supremacy and the need for organizations led by black people." He thought it particularly important that URPJ allowed him to "get access to information on white supremacy . . . and solution-based information." Through his URPJ work, he extended his activism to include community-based economics such as time banks and cooperatives. Race and racism, he stated, is a by-product of economics; thus, "a big part of our struggle is to control economics in a community." With URPJ's support, he is helping to establish a black think tank to provide support for formerly incarcerated men who are reentering society and to foster healthy relationships and empowerment.

Another interviewee was a middle-aged white woman who had lived in Detroit for fifteen years. She came into URPJ with experience in antiracism training through the Michigan Roundtable for Inclusion and Diversity. She became motivated to work with URPJ because she saw "how many African Americans are forced to live due to conscious decisions and self-centered concerns [of whites] that, collectively, had social-level impacts." She came to understand "the material consequences of white flight and the struggles of black people."

This woman uses the lessons of URPJ in her day-to-day interactions with those in her neighborhood, where she tries "to be intentional about addressing racism" whenever she perceives it. She also uses lessons from the power analysis training to "consciously examine power in her neighborhood organization"—confronting members with any potential racism or racial exclusion within the group. She believed that participation in the White Caucus was an important way for her and other whites to hold themselves accountable in their antiracism work. In an interview at her home, she was asked how she applies the lesson of unlearning racism. She replied that in one of the many Detroit neighborhoods that people are attempting to revitalize through gardening and urban farms, adopting a simple lifestyle was one of the biggest things to promote.

She actively engages children in gardening, and within this context she is receptive to issues of race and racism raised by the children. Her conversations are grounded in efforts to question and challenge racism. She does this by supporting the children's agency while acknowledging the reality of their limited social power. She also promotes a positive view of blackness in the garden, countering the prevailing negative views of darkness and blackness in our society. In her conversations with young gardeners, she finds ways to communicate "that black is beautiful, and valuing that." For example, she says to the children, "The darker the earth is, the better it is. Now we have this black soil, and it's best when it's really, really dark."

SUMMARY

Within Detroit's long history of progress for social and economic equality, URPJ members are embarking on a new journey to define and remove racism in a way not done before. Whether it is working closely to help an African American–owned grocery store remain open or empowering its members to probe deeply to pursue their own work on racism, URPJ is a fluid, powerful organization that owes as much to its principles and theory of change as to its humanistic approach to catalytic change. URPJ believes that change best occurs through meaningful, honest, and open conversations about race within the specific contexts and settings where people live and work, and by developing conscious relationships through these conversations.

The example of the middle-aged white woman—her expressions of everyday practices, reflections, and insights—affirms this theory of change. It is within the context of her everyday life, with the people she encounters, that she applies critical knowledge and has meaningful conversations about race that challenge conventional and largely negative views of her neighborhood and black people. The hope is that her story, and those of other URPJ members, will inspire others to apply critical knowledge to meaningful conversations about race, especially as they relate to food security and food sovereignty. Over time this will lead to the vision of a just food system that establishes healthy, fresh, and affordable food as a fundamental human right and in which personal, institutional, and systemic racism have been dismantled.

The authors thank Malik Yakini and Lila Cabbil for their contributions to this chapter.

CHAPTER TEN

Achieving Community Food Security through Collective Impact

GREG LAWLESS, STEPHANIE CALLOWAY,
AND ANGELA ALLEN

In this chapter Greg Lawless, his Cooperative Extension Service colleague Angela Allen, and Stephanie Calloway introduce the model approach of collective impact and its applicability in food system practice. They describe how the Milwaukee Food Council is using this concept to achieve community food security. This is a CRFS value oriented to strong communities and to system thinking and collaboration.

UNDERSTANDING COLLECTIVE IMPACT

Simultaneous with the start of the CRFS project, the *Stanford Social Innovation Review* published an article called "Collective Impact" by John Kania and Mark Kramer (2011) of the consulting firm FSG. Their model of collaboration is for a group of important individuals from different sectors to commit to an agenda for solving a specific social problem.

The model reflects the systems approach of its practitioners. To tackle a persistent crisis of student achievement in the Greater Cincinnati area, which the authors described in their sample case study, the leaders realized that to effectively fix one point on the educational continuum, all parts of the continuum must improve at the same time (Kania and Kramer 2011). The participants also understood that complex and urgent problems like disparities in education, polluted watersheds, or childhood obesity cannot be solved by governmental, business, academic, and nonprofit organizations working independently or at odds.

Kania and Kramer (2011) clearly stated collective impact's five guiding concepts:

- **A long-term commitment.** This means years of patient, adaptive efforts and continuous financial support. Funders must be willing to let grantees determine the direction of the work.
- **Important actors.** They could include influential voices in the community as well as people who manage businesses and control resources that affect community food security.
- **The involvement of multiple sectors.** Although the authors noted that this is a monumental challenge, it is central to collective impact. It is not enough to get nonprofit organizations and university specialists working together—especially in the food system, where private businesses play such a major role.
- **The development of a common agenda.** This is one of the authors' five conditions of collective success and is considered below.
- **Solving a specific social problem.** This is significant in two respects. First, the intention to solve a problem—not simply to address or alleviate it—is bold and explains why serious commitment is required from every sector. Second, it is a challenge to reach consensus on which aspect of a multifaceted problem takes the highest priority.

This foundational article impressed the CRFS project team, which distributed it to the project's growing group of community partners across the country. These partners had frequently discovered the article on their own—national interest in the concept had spread quickly. In Wisconsin, the Milwaukee Food Council (MFC) made a commitment to put the model into action the same year it was introduced by the FSG consultants. The MFC's experience is the focus of this chapter.

The following sections introduce twenty distinct concepts that appeared in four seminal collective impact articles from 2011 to 2015. These concepts assess both the model itself and the Milwaukee attempt to eliminate food insecurity by putting the model into practice.

Three Preconditions for Collective Impact

Before reviewing the more familiar five conditions, it is instructive to consider the following three preconditions for collective impact, which were introduced by Hanleybrown, Kania, and Kramer (2012):

- **The recruitment of an influential champion.** This person must command the necessary respect to be able to bring CEO-level cross-sector leaders together and sustain their engagement over time.
- **Adequate financial resources.** These must last for at least two years, and there should be at least one anchor funder—someone who is engaged from the beginning and who will mobilize other resources.
- **A sense of urgency.** This brings together important actors who may be disconnected or in competition for funding and attention. Has a crisis created a breaking point to convince people that an entirely new approach is necessary?

Five Conditions for Collective Success

The preconditions set the stage for the following five conditions for collective success, which were introduced in 2011 and expanded upon in subsequent articles:

- **A common agenda.** This should create the boundaries of the problem to be solved and a guiding strategic action framework, which includes a description of the problem informed by solid research, a clear goal for the desired change, a portfolio of key strategies to drive large-scale change, a set of principles that guide the group's behavior, and an approach to evaluation that will obtain and judge the feedback on its efforts (Hanleybrown et al. 2012).
- **A shared measurement system.** These should be present at the community level and throughout all participating organizations to ensure that all efforts remain aligned (Kania and Kramer 2011). However, competing priorities among the participating organizations and their fear of being judged as underperforming make it difficult to reach agreement on common measures (Hanleybrown et al. 2012).
- **A backbone support organization.** This must have a small, dedicated staff that is separate from the participating organizations and that provides facilitation, technology and communications support, data collection and reporting, and logistical and administrative support (Kania and Kramer 2011).
- **Mutually reinforcing activities.** These should build on existing efforts in the community. Such activities are not managed centrally by the backbone organization but involve cascading levels of linked collaboration with the stakeholders through working groups and partner organizations (Hanleybrown et al. 2012).

- **Continuous communication.** This is necessary to keep all stakeholders involved, informed, and contributing to the collection of data to measure impact. The backbone organization is largely responsible for facilitating this communication and providing periodic assessments of the progress of the work groups (Hanleybrown et al. 2012).

Embracing Emergence

A third article (Kania and Kramer 2013) emphasized the need for flexibility and discouraged predetermined solutions. Developed from the field of complexity science, the concept of emergence describes events that are unpredictable, that appear to result from the interactions of elements, and that no organization or individual can control.

In this context of emergence, the authors introduced two adaptive concepts. The first, *collective vigilance*, explains how funding deficiencies and other roadblocks can be overcome when organizations look beyond their particular agendas and discover existing resources and emerging opportunities they would otherwise miss. The second, *collective learning*, is a flexible and developmental evaluation process that revises goals and strategies in response to the indicators used to measure progress and impact, as well as changes in the broader environment, the systems of interaction, and the capacities of participants (Kania and Kramer 2013). This consideration of changing circumstances monitors relationships and conditions and reveals new resources and potential solutions. It requires the leaders of collective impact initiatives to embrace uncertainty and requires a shift in the mindset of funders to support processes that lead to emergent solutions.

Civic Culture and Community

The aforementioned articles make a persuasive case for collective impact as a theory of change, but as people across the country put the theory into practice, some began to recognize deficiencies. Paul Schmitz (2014), the former chief executive officer of Public Allies and a senior advisor to the Collective Impact Forum, wrote that "much of the early research and work on collective impact has emphasized the structural, strategic, and measurable. To succeed long-term, there must be more attention paid to the cultural."

The Collective Impact Forum (2014) published several essays by field practitioners that included highlights from a roundtable on community engagement and collective impact. One roundtable participant, Richard Harwood, claimed there was often too little involvement by grassroots community members and too much dominance by specialists and professionals from outside the community whom Harwood and others called the "grasstops." He warned that "the danger of grasstop power is that . . . people see a group of professionals in their community who have dreamed something up" that does not engage or benefit them.

The Collective Impact Forum invited Harwood to expand on his comments in a separate article (Harwood 2015). In it he presented five characteristics of a community's civic culture that determine whether a collective impact effort will succeed or fail. Taken together, these characteristics "explain why some communities move forward and others remain stuck or treading water; and why some communities that do make progress ultimately slide backward." The five characteristics are as follows:

- **Ownership by the community.** This means that no matter how many "important actors" and "influential champions" you assemble to create a common agenda, the impact will not be sustained if the community residents are not genuinely engaged in the effort.
- **Strategies that fit the community.** These recognize what stage a community is moving through as its civic culture grows or declines. The stages involve the degree of agreement and connections among the residents and their self-perception as a community.
- **A sustainable enabling environment.** This can be assessed by several measures, including social gatherings and spaces for interaction, boundary-spanning organizations, safe havens for community discussions and decision making, and community norms that guide how people interact.
- **A community's belief in itself.** This means that the residents develop the confidence that they can get something done together. This can involve small but successful actions in which the residents are active participants and leaders.
- **A community's narrative.** This derives from the accumulation of stories that the residents tell themselves and others about how the community deals with its challenges. A can-do narrative develops when the members experience repeated success through collective action.

The next section uses the characteristics of civic culture listed above and other distinct collective impact concepts to evaluate an initiative to alleviate food insecurity in Milwaukee.

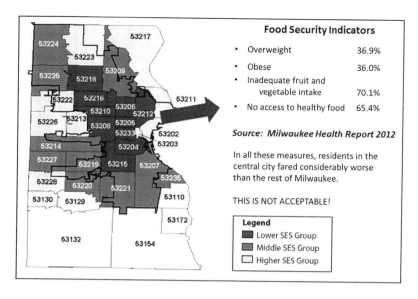

Food Security Indicators

- Overweight   36.9%
- Obese   36.0%
- Inadequate fruit and
  vegetable intake   70.1%
- No access to healthy food   65.4%

Source: Milwaukee Health Report 2012

In all these measures, residents in the central city fared considerably worse than the rest of Milwaukee.

THIS IS NOT ACCEPTABLE!

Legend
- Lower SES Group
- Middle SES Group
- Higher SES Group

FIGURE 19. Food security data for lower socioeconomic status zipcodes from the Center for Urban Population Health used in the MFC's collective impact interview guide. (Courtesy of Center for Urban Population Health, Milwaukee)

THE MILWAUKEE FOOD COUNCIL'S COLLECTIVE IMPACT INITIATIVE

The city of Milwaukee has about six hundred thousand residents. CRFS research was focused primarily on the ten contiguous zip codes that the Center for Urban Population Health reports as representing lower socioeconomic status (SES) populations, compared with the rest of the city (fig. 19).

In 2011 these ten zip codes represented roughly half of Milwaukee's population. Some areas were especially segregated and economically challenged. For instance, just south of Interstate 94 in the 53204 zip code, 65 percent of the residents were Hispanic, the median household income was about $19,000 per year, the area had 17 percent unemployment, and 39 percent of the households received food stamps. North of I-94 in the 53206 zip code, 97 percent of the residents were African American, the median household income was just over $22,000 per year, the area had 31 percent unemployment, and 46 percent of the households received food stamps (US Bureau of the Census 2011).

Tracking a limited set of food-related measures, the center reported that 36.9 percent of the people in this area were overweight, 36 percent were obese, and 70.1 percent were not eating adequate fruits and vegetables. By measuring access to healthy food in terms of supermarket concentration, the lower SES areas had considerably less access than wealthier parts of the city (Greer et al. 2013).

These statistics provide an incomplete summary of the contributing factors and associated health outcomes of food insecurity in Milwaukee. Finding better ways to understand and address food insecurity has been an ongoing concern of the MFC since it was established in 2007. The group was originally convened and supported by Martha Davis Kipcak, who was then employed as a community food activist by a local philanthropic foundation. In a 2015 interview she said, "The initial vision was to create a common table open to all stakeholders to work together toward a healthy, sustainable, equitable, just, and vibrant food system in Milwaukee. Inclusive, collaborative, thoughtful, and meaningful work has been central to the efforts, resulting in slow but steady systemic changes for Milwaukee's food community."

As an informal, volunteer-run network, the MFC decided not to use the word *policy* in its name because it did not want to limit itself only to policy development or advocacy. Also, even though participation by government agency staff members was always welcome, there was a strong agreement to remain independent from city or county government. (See chapter 12 for more on the general characteristics of food policy councils.)

As the network expanded, its participants began to ask whether and how the MFC could support activities outside the bimonthly meetings to address urgent food-related issues and to develop opportunities. In 2011 the group completed a strategic planning process that outlined six "commitments meant to develop into specific goals." These goals related to policy change and advocacy, equity and justice, respect for diverse food cultures, economic vibrancy, ecological sustainability, and, at the top of the list, collective impact.

In 2012 one MFC participant, the executive director of a backyard gardening organization, led a metrics subcommittee to work toward one of the five conditions for collective success: a shared measurement system. Initially some university collaborators were approached to help identify existing sets of data that could cost-effectively track progress toward the council's food system goals. However, it soon became clear that those goals were not fully developed.

The subcommittee turned its attention to restating the six commitments of the council's earlier strategic document as a set of goals. It then identified examples of specific outcomes representing progress toward those goals. In other words, what would success look like? Under the goal of ecological sustainability, for example, the subcommittee listed the decreased use of fossil fuels and the increased use of water-conserving techniques at all food-related venues. Under another goal—to develop a food system that nurtures the physical, mental, and spiritual health and wellness of all residents—it listed a reduction in food-related deaths and physical illnesses in Milwaukee County and an increased number of MFC members representing faith-based organizations. The subcommittee shared drafts of this work at the MFC's bimonthly meetings. They had good intentions to identify data sources for metrics that could track progress toward all of the stated goals and outcomes. However, when summer gardening and farming activities resumed in 2013, the subcommittee work was postponed.

In January 2014 a session on collective impact at the eighth annual Wisconsin Local Food Summit inspired Stephanie Calloway, the garden and nutrition coordinator for a nonprofit health and wellness organization on the city's south side, to propose reestablishing an MFC subcommittee to pick up where the strategic thinking process and metrics subcommittee left off. The first phase of that effort is described in the following section.

Launching the Collective Impact Initiative

Despite the available articles on collective impact, the subcommittee struggled to find a clear path to develop an initiative. Which of the five conditions for collective success should come first: a shared measurement system, mutually reinforcing activities, or a common agenda? The literature was not clear on this point.

However, the collective impact authors consistently cautioned against jumping to solutions too early. In fact, they said that developing a common agenda was not about creating predetermined solutions (Kania and Kramer 2013). If starting with solutions was not the answer, the subcommittee decided, it might make sense to start with the problem.

At the MFC meeting in March 2014, about twenty people expressed an interest in developing the initiative. From this group, eleven subcommittee members met in April to begin working on the problem, using definitions and references from the collective impact literature. Although the conversation encompassed a wide range of topics, the members were reminded that the authors recommend focusing on one urgent problem to solve.

The conversation evolved into two potential themes. The first was food insecurity, described informally as "not enough people in Milwaukee are eating enough good food." The second was about environmental sustainability, which was described as "our current agricultural system is driving us over an ecological cliff."

The subcommittee took these two ideas back to the full council in May 2014 and facilitated small-group discussions to determine which of these two themes resonated or if other issues were more urgent. Overwhelmingly, the council members agreed that food insecurity was the most pressing issue, although the issues of racism and insufficient diversity in the MFC's membership were brought up multiple times.

With this feedback and approval from the MFC membership to focus on food insecurity, the subcommittee believed that more information was necessary to understand and define the problem. To gain greater community feedback, the subcommittee created an interview guide that introduced the MFC's collective impact initiative to tackle the problem of food insecurity. A key survey question in the guide was "What is your vision for a food-secure Milwaukee (or what would it look like if enough people in Milwaukee were eating enough good food)?"

As the collective impact authors warned, the urge to jump to solutions is strong; the subcommittee believed that this question could illuminate the goals and focus areas as it defined the scope of the initiatives while appeasing the human inclination for problem solving. The interview guide also asked the participants to define what "good food" means to them—a question that would guide the development of specific metrics. The remaining survey questions gauged the respondents' willingness and ability to commit to the initiative and solicited names of other people to engage.

The primary vehicles for survey completion were a SurveyMonkey link distributed through Facebook, the MFC listserv of more than three hundred members, and personal e-mail distribution. The survey was also shared with community groups, at bus stops, and outside the county social services building. A total of 174 surveys were collected, and the responses were organized into statements about goals, strategy, and vision. Goal statements represented results or achievements to which a collective impact initiative would be directed. Strategic statements defined plans or actions to achieve the goals. Vision statements represented thinking about the future with imagination and wisdom.

Eight themes emerged from the survey data; after some rewording, these themes were presented as follows at the MFC meeting in November 2014 in rank order across all survey audiences:

1. **Improving food environments.** Accessibility, more grocery stores, farmers' markets, healthy food options, convenience, zoning laws, and fresh vegetables in school lunches.

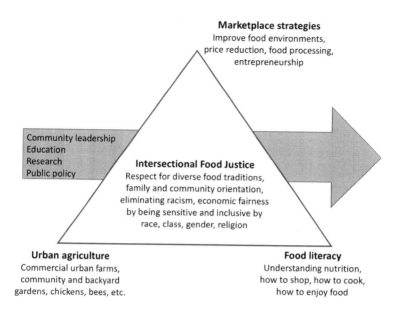

Marketplace strategies
Improve food environments,
price reduction, food processing,
entrepreneurship

Community leadership
Education
Research
Public policy

Intersectional Food Justice
Respect for diverse food traditions,
family and community orientation,
eliminating racism, economic fairness
by being sensitive and inclusive by
race, class, gender, religion

Urban agriculture
Commercial urban farms,
community and backyard
gardens, chickens, bees, etc.

Food literacy
Understanding nutrition,
how to shop, how to cook,
how to enjoy food

FIGURE 20. A strategic framework for addressing food insecurity, developed in 2015 by the MFC. (Courtesy of Stephanie Calloway, on behalf of MFC)

2. **Incorporating good food and community life.** Value of food, family and friends, community involvement, and culturally viable food.

3. **Promoting food justice.** Race, class, sex, religion, food deserts, and lower obesity rates.

4. **Improving food quality.** Non–genetically modified; fewer or no chemicals; fresh, nutritious, healthy food.

5. **Making good food more affordable.** Cheap fresh fruits and vegetables, low-cost organic foods, ability to use EBT cards at farmers' markets.

6. **Fostering economic development.** Jobs, elimination of poverty, improved transportation.

7. **Developing educational strategies.** Cooking and nutrition classes, informed consumers, awareness.

8. **Growing more food in the city.** More gardens, education on growing food, access to growing food.

Many attendees said it was difficult to prioritize these issues because they are so interconnected. Through additional conversations at the bimonthly MFC meetings, the eight themes were sorted into three categories: marketplace strategies, food literacy, and urban agriculture (fig. 20). Despite the low ranking of "growing more food in the city," MFC members agreed to maintain a prominent space for urban agriculture in their strategy because of the important work being done in that area.

The MFC also listed four types of general activities that cut across all three strategies: research, education, public policy, and community development, as shown in the figure's gray arrow. One promising development along these lines is the formation of the Cooperative Institute for Urban Agriculture and Nutrition (CIUAN, pronounced "chew on") to bring together academic institutions in Milwaukee and Madison, the city government of Milwaukee, Growing Power, and MFC members. With support from CIUAN, the MFC hopes to increase community input, refine its goals, and determine the best ways to measure improvements in food security in Milwaukee. The MFC also hopes to continue asking community residents to define *good food*. The responses to this question may help the group focus its efforts to build a shared system of metrics and shape a common agenda.

Finally, the MFC adopted the concept of intersectional food justice as central to everything it does. This approach to addressing food insecurity in Milwaukee was presented to the MFC in January 2015, with the proposal to develop three- to five-year plans for each of the triangle's corners in figure 20, and each plan was to be grounded in an intersectional, antioppression framework.

Discussion centered on the importance of including marketplace strategies, food literacy, and urban agriculture in any initiative to ensure sustainability and the overall strength of the system. Future steps include holding a workshop on undoing the racist practices that act upon the Milwaukee food system and mapping current marketplace activities and assets in the food environment.

The subcommittee will continue to incorporate information from surveys and follow-up MFC meetings with the strategic thinking document and metrics framework from previous years. The MFC is working with community partners to develop ideas and strategies for bringing this information into the community for greater input and investment.

Update on the Collective Impact Initiative

The previous section covered the activities in 2014 that culminated in a preliminary strategy presented at the statewide local food conference that had inspired Stephanie

Calloway one year earlier. Under her continued leadership the MFC proceeded with a series of joint meetings with CIUAN in 2015. The latter had established the Community Collaboration Council to help develop its research and education agenda, and there was a promising overlap of interests.

Later the Collective Impact Forum (2015) offered a three-part webinar series on getting started with collective impact. The webinars brought MFC and CIUAN members up to speed on the collective impact model approach and raised important new questions, such as how to design a decision-making and management structure for an initiative involving multiple organizations and sectors.

Through the winter and spring of 2016, Calloway arranged multiple meetings and video conferences with other MFC members to tighten and clarify the boundaries of the initiative. A new approach focused on making healthy food more accessible geographically, more affordable, and more desirable or appealing to diverse populations in the city. These three goals were used to re-evaluate the metrics work from 2012.

Continuing to build its infrastructure as a collective impact backbone organization, the MFC filed for nonprofit status as a member organization in 2016. It continues to operate without paid staff. When the first board president left, Calloway took over the role.

A draft document that the collective impact work group shared with MFC members in July 2016, entitled "Good Food for All," listed the following five initial focus areas:

- Promoting programs that work, involving media, nutrition education, and story sharing
- Economic development, food processing, distribution, and marketing
- Food access, focusing on chronic food shortage
- Local food production, urban farms and gardens, and composting
- Convening food system stakeholders and mapping food system activities

Building on previous work identifying metrics to record food system change, the MFC collective impact work group developed a matrix for action (strategic thinking area, desired outcomes, associated metrics, and associated strategies). Using the matrix and this strategic plan, the principles and practices of collective impact will be extended to other MFC work groups. The collective impact goals will infuse the other work group activities and products; reciprocally, the other work groups will help refine and expand the focus areas described above.

TABLE 11. Assessing the MFC through Collective Impact Concepts

Collective Impact Concepts	Assessment of MFC's Collective Impact Initiative
Guiding Concepts (Kania and Kramer 2011)	
Long-term commitment	Alignment
Important actors	Potential alignment
Multiple sectors	Potential alignment
Common agenda	Potential alignment
Specific social problem	Alignment
Preconditions (Hanleybrown et al. 2012)	
Influential champion	Potential alignment
Adequate financial resources	Divergence
Sense of urgency	Potential alignment
Conditions for Collective Success (Kania and Kramer 2011)	
Common agenda	Potential alignment
Shared measurement system	Potential alignment
Backbone support organization	Potential alignment
Mutually reinforcing activities	Potential alignment
Continuous communication	Alignment
Embracing Emergence (Kania and Kramer 2013)	
Collective vigilance	Potential alignment
Collective learning	Alignment
Civic Culture (Harwood 2015)	
Ownership by the community	Potential Alignment
Strategies that fit the community	Potential alignment
Sustainable enabling environment	Alignment
Community's belief in itself	Potential alignment
Community's narrative	Potential alignment

The earlier discussion of collective impact presented twenty concepts in four articles published from 2011 to 2015. As part of the CRFS assessment process, these concepts were used to characterize the MFC's efforts to employ collective impact to eliminate food insecurity in Milwaukee.

The left column in table 11 groups the collective impact concepts as they appeared in the literature and as they have been listed above. The right column provides an approximate characterization of the MFC's alignment with each concept. These represent the authors' subjective perspective as of September 2016. Some assessments were difficult to make and imprecise, and several have changed over time toward stronger alignment.

MFC members have shown a continuous interest in collective impact since 2011, and their consistent effort to launch an initiative since early 2014 demonstrates a multiyear commitment. They have narrowed the specific social problem to food insecurity in Milwaukee, an urgent issue for every resident who suffers its consequences.

The MFC has spent much of its energy in the past three years developing a common agenda. The "Good Food for All" document that was drafted and shared with the membership in July 2016 represents a significant step toward a strategic action framework, but it still needs work. The social, economic, and public health consequences of food insecurity in Milwaukee must be clearly presented and informed by solid research. The document's portfolio of five focus areas will require more specific strategies and methods for evaluation. A set of principles to guide the behavior of the collective impact initiative's participants might also be useful.

The MFC has made notable progress on the collective impact approach without a paid internal staff. Some of the work was done by people compensated by other entities, sometimes through complimentary grant projects. This overlapping or alignment of a collective impact initiative with the missions and commitments of participating organizations and the combination of resources is a fundamental reason to collaborate.

However, it is clear that the MFC cannot serve as the backbone organization that is required to coordinate a strategy to eliminate food insecurity in Milwaukee unless it can secure more substantial financial resources. This is the MFC's only clear divergence from the collective impact model to date.

The MFC board of directors, recognizing the need to acquire more funding, has chartered as a Wisconsin nonprofit corporation and has gained 501(c)(3) tax status. Meanwhile, several collective impact work group participants are looking for grant sources not normally available to community organizations, like federal research funding. It should also be possible to assemble more resources, including staff time, from participating organizations to support the Good Food for All initiative.

One promising approach involves greater integration with, or a partnership agreement between, the MFC and CIUAN. For several years their relationship has been informal and sometimes ambiguous, but there appears to be strong interest in both organizations in expanding their collaboration. An area of work that academic associates of CIUAN might help coordinate is the development of meaningful and appropriate metrics that participating organizations can track and contribute toward a shared system. CIUAN may also be able to mobilize university students to support the collective impact initiative in fulfillment of their course work or degree requirements.

One of the webinars offered by the Collective Impact Forum in 2015 broached the subject of governance structures for collective impact initiatives. However, the four seminal articles described earlier offer little guidance on this critical issue. Any further collaboration with CIUAN will require much more clarity about who makes decisions and how shared resources are managed.

Any new partners or funders in the MFC's Good Food for All initiative will also want to know how leadership and management of resources will be handled. Currently the MFC membership is largely composed of representatives from non-profit organizations, public agencies, and academic institutions. Because food- and health-related companies play such an essential role in Milwaukee's food system, more small-business owners and corporate representatives must be engaged.

Although Calloway's leadership during the past three years has been critical, it may be necessary to attract "important actors" and an "influential champion" to raise funds that will be necessary to support the collective impact initiative for several years. This raises a significant challenge involving the tension between the grassroots and "grasstops" levels of participation introduced earlier in the chapter. The first collective impact articles clearly called for real and consistent involvement of "CEO-level cross-sector leaders." These leaders control significant public and private resources that can be used to implement the common agenda and strategic action framework. Several nonprofit executive directors attend MFC bimonthly meetings, but none currently serve in the collective impact work group, nor do any public agency staff members or business representatives.

It is worth asking why MFC and its collective impact efforts have not attracted more grasstops participation. Within the food policy council literature, Schiff (2008) and Harper et al. (2009) acknowledge that most councils are still developing the credibility to influence local government policy. In the latter report, the authors and leaders quoted mention the need to build credibility more than ten times. Despite the steady development of the MFC since 2007, it still has to persuade more leaders in government, business, academia, and the funding community that its priorities and activities merit their support and involvement.

At the same time, it might not be accurate to call the MFC itself a grassroots organization. Many of its active members work in professional capacities for entities that share the MFC's priorities. Many do work directly with populations experiencing food insecurity. However, stronger efforts will be necessary to engage people from those communities in the planning and implementation of the Good Food for All initiative.

Harwood (2015) emphasized this need for genuine community engagement, and since 2007 the MFC has made considerable progress toward a "sustainable enabling environment" by creating a bimonthly space for social interaction, community discussions, and democratic decision making. In time the council may develop a true spirit of community ownership in the collective impact initiative if it can make good food affordable, accessible, and desirable in ways that fit the communities it intends to serve.

One promising aspect of the MFC's Good Food for All initiative is the strong commitment to racial and cultural inclusion, along with an acknowledgment of the systems of oppression that contribute to food insecurity. Most of the collective impact literature has not directly addressed issues of racism, white privilege, or disparities of income and power that make collaboration in a large city like Milwaukee so challenging. The MFC can inspire and guide other cities if it demonstrates how intersectional food justice can overcome obstacles to achieving community food security. A longitudinal assessment of the Good Food for All initiative would be useful in determining whether the model approach of collective impact described in this chapter can provide an effective means to develop community and regional food systems.

CHAPTER ELEVEN

Education and Food System Change

DESIRÉ SMITH AND STEVE VENTURA

In this chapter Steve Ventura introduces Desiré Smith, a recent University of Wisconsin–Madison graduate. She describes her experience as the coordinator of a UW–Madison program to encourage youth from underserved communities to consider a pathway to college and food system careers.

Whenever I (Steve) got to know people involved in food system change from various cities, I often asked them this question: "Can this food system activity engage young black men in positive activities in your community?" The subtext to the question was that young male African Americans in poor neighborhoods suffer disproportionately high rates of unemployment, incarceration, violence, and, at least from an outside perspective, despair and disconnection. I asked with the hope that efforts to improve food security and food justice could address some of these broader social ills.

My question was often met with a sigh and an answer that said, in effect, "No, once they are on the path of drugs, guns, gangs, and prison, there's little hope that food systems will be of interest." Glimmers of hope have emerged, however, through such programs as Project Return in Milwaukee, which teaches parolees to garden. As Executive Director Wendell Hruska explained, these men have a hard time getting jobs, so producing food is one way they can feel self-worth, by helping to support themselves and their families and by learning the value of applying themselves to a task. Some prisons are revitalizing work farms, such as the minimum-security correctional institution in the village of Oregon, Wisconsin, where prisoners grow sixty thousand pounds of vegetables for their own consumption annually.

A follow-up suggestion to my question was common among food system activists in every community, whether black, white, Hispanic, poor, or middle income: "Children need to learn about food systems. We need to engage them when they're young." This mantra came up not just in relation to young black men; it was universal. It came up in reaction to the alleged belief that growing food was "stoop labor" left behind in the South or in another country and thus was something demeaning. It came up in discussions of the prevalence of obesity and related health issues. It came up in discussions of school meals programs. It came up in discussions of career paths and the value of STEM (science, technology, engineering, and math) curricula.

The Social Justice Learning Institute (n.d), an affiliate in the CRFS project, works with some of the neediest communities in Inglewood (California) and Los Angeles, and it emphasizes health equity and prosperity through its food system programming: "We know that a quality education is foundational to a prosperous future, and for that reason, education is at our core. We specialize in culturally relevant learning, teaching, and curriculum development, which supports positive identity growth, increases academic competencies, and expands opportunities for civic participation."

Many dynamic programs, activities, and curricula have been created to help children understand what food is, where it comes from, what it does for our bodies, and why it's important to know about these things. Entire books have been written about the value of experiential learning through activities such as school gardens. Farm-to-school programs are helping the next generation understand where its food comes from and who produces it. Several CRFS Innovation Fund projects included youth education. Many project partners have youth internship programs; in some cases, this is a substantial focus of their activities (see section).

As the recipient of a USDA-distributed grant, the CRFS project was required to have research, outreach (Cooperative Extension Service participation), and educational components. The educational activities included typical college-level opportunities such as undergraduate internships, graduate student support, and seminars. Fortunately, we were also able to connect with a pre-college program at the University of Wisconsin–Madison. The Pre-College Enrichment Opportunity Program for Learning Excellence, or PEOPLE, was developed to help close the gaps in educational opportunities and attainment for minorities and to help them think about college as a viable post–secondary school option. It is designed for disadvantaged Wisconsin students with strong academic potential, and particularly for students who are the first in their families to attend college.

Our project supported the development and delivery of one-week and three-week urban agriculture curricula for incoming high school sophomores and juniors in the summer program, and it provided internships for incoming seniors. The final portion of this chapter is a description of PEOPLE from the perspective of one of the UW–Madison students we hired to help organize the summer programs: Desiré Smith.

Thanks in part to our incredibly positive experiences with the project interns who helped organize PEOPLE—particularly Desiré and another African American student, George Reistad—we recognized the need for and value of leadership development. Remarkably and fortunately, George moved on to a position as the assistant policy director at the Michael Fields Agricultural Institute in East Troy, Wisconsin. He now serves as the food policy coordinator of the City of Madison.

For several years, George's boss at Michael Fields, Margaret Krome, has had a highly successful intern program for developing leaders in sustainable agriculture. She recruits primarily at the postgraduate level, and several interns have gone on to influential positions in the sustainable agriculture movement. In spite of her considerable effort, it has been very difficult for her to find viable candidates who are not white middle-class women. It was apparent to George and Margaret that part of the answer was to develop a complete pipeline: to interest students in studies and careers in food systems and/or sustainable agriculture at an early age and, through a variety of mechanisms, provide continuous support from high school or even earlier through postcollege internships and placements. We believe that PEOPLE will become an important component of a pipeline to leadership positions.

Community GroundWorks Orchards Project: A Different Kind of Common Core

SHELLY STROM

Fruit trees transcend differences of age, sex, race, religion, culture, sexual orientation, and zip codes. They also create common ground through shared stewardship and foster hope for the future. Like community gardens, community orchards are innovative projects that strengthen the local food system and improve the social and natural environment. However, orchards rely on deep collaboration and long-term commitment, because it may take years for trees to produce ripe fruit. Before the harvest, communities can reap the social benefits nurtured through public orchards.

Thanks to an Innovation Fund project grant from the CRFS project, Community GroundWorks planted hundreds of fruit and nut trees in Madison in 2014 and 2015, with local partners at public schools, neighborhood centers, and community gardens. Community GroundWorks is a Madison-based nonprofit organization that connects people to nature and local food. Community GroundWorks serves diverse communities at Troy Gardens, its home site of twenty-six urban acres and at schools, community centers, and community gardens throughout Dane County and beyond. Through hands-on education, children and adults learn about gardening, urban farming, healthy eating, and environmental stewardship. The Community Orchards Project put organizational values into practice and demonstrated our vision of "growing food, growing minds, together."

The original intent of the Community Orchards Project was to create a handful of small urban orchards where interest and need intersected. To support socially just and fair access to healthy food, the Community GroundWorks staff sought out and cultivated relationships with underrepresented populations. A key objective was to create a series of models to inform future projects in a wide range of locations: schoolyards, public parks, apartment complexes, food gardens, and green spaces.

To amplify its influence, the Community GroundWorks staff creatively integrated the orchard project with other organizational initiatives; fruit trees became the perfect invitation for community engagement, conversation, and envisioning. The Community Orchards Project supported seventeen schools, six community gardens, and two neighborhood centers; seven thousand people were directly involved in the planting and primary care of more than 550 trees. The project included seasonal educational workshops, consultation, orchard design and installation, tree-planting events, and community celebrations like Pi(e) Day, Mulberry Fest, and a Harvest Party. The project leveraged additional grant support, and more orchards were under way in 2016.

One of the biggest events of the Community Orchards Project took place in May 2015, when Sandburg Elementary School on Madison's north side held a tree assembly in honor of its new community orchard (fig. 21). All 453 students, along with teachers, parents, neighbors, and government officials, gathered outside the school to sing songs about trees, recite tree poetry, and celebrate the cherry, plum, apple, and pear trees planted by the students.

FIGURE 21. Students plant a fruit tree at Sandburg Elementary School in Madison, Wisconsin, May 2015. (Courtesy of Community GroundWorks)

Throughout the duration of the project, everyone had stories about their favorite fruits and about eating apples, pears, and mulberries straight off the trees. People shared their memories of fresh and local food, creating a "common core" of social and cultural experiences. In the words of a kindergarten-age orchardist upon discovering a shared love of cherry trees, "We all have so much in common!"

PEOPLE: THE PRE-COLLEGE ENRICHMENT OPPORTUNITY PROGRAM FOR LEARNING EXCELLENCE

I (Desiré) was able to work with PEOPLE and the CRFS project in researching and participating in ventures to create an urban agriculture curriculum for high school students. This curriculum seeks to expand students' knowledge of the importance of engaging in the food system while offering them opportunities to explore possible career paths. After four years of implementation, the urban agriculture program seems to have a bright future; with that in mind, the CRFS project has begun implementing strategies to expand the program and better serve PEOPLE students. The program is dedicated to helping "students successfully make each transition from middle school to high school and from high school to college," and the food security and food systems theme seems to be a good ladder to success. (All quotes are from the program literature.)

PEOPLE is one of the nation's most comprehensive diversity scholarship pipelines for developing talent among underrepresented, economically disadvantaged, and first-generation university students. The primary goals of the program are to do the following:

- Increase the number of Wisconsin high school graduates who apply, are accepted, and enroll at UW system institutions.
- Encourage partnerships that build the educational pipeline by reaching children and their parents at a child's early age.
- Increase access for underrepresented minorities and women, especially in STEM.
- Eliminate the achievement gap between majority and underrepresented students.

PEOPLE accepts highly motivated students into a rigorous program to build study skills as well as interpersonal and communication skills, explore and strengthen academic and career interests, and gain a positive experience on a world-class campus. PEOPLE is definitely UW–Madison's most successful venture in creating such opportunities and improving campus diversity. More than half of PEOPLE's high school students are admitted to UW–Madison with full-tuition scholarships. Once enrolled in the university, PEOPLE students maintain about a 90 percent retention rate. All PEOPLE high school scholars graduate from high school, and 94 percent enroll in higher-education institutions. In comparison, in 2014 in the Madison Metropolitan School District, 56 percent of African Americans and 60 percent of free or reduced-cost lunch students (an indication of low-income families) completed high school in the standard four years. The figures are similar in Milwaukee, which makes up a substantial portion of PEOPLE participants.

From its inception, the CRFS project established a partnership with PEOPLE. The rest of this chapter discusses PEOPLE's urban agriculture curriculum and how the CRFS project integrated this new educational track into the well-established structure of PEOPLE's high school summer program.

To understand how the urban agriculture curriculum was incorporated into PEOPLE, it is important to understand the structure of the entire program and the holistic value it places on supporting and guiding students on their journey through primary, secondary, and higher education.

PEOPLE consists of the following four separate, chronological programs:

- Prep program (elementary school program)
- Middle school scholars program
- High school scholars program
- College scholars program

Students in the PEOPLE prep and middle school programs are provided with early exposure to potential college majors and career options, after-school tutorial services in core subject areas (math, science, English, history, and world languages), academic and cultural enrichment services, leadership development, informal mentoring, and other college preparatory services. PEOPLE partners with the various UW schools and colleges to provide academically grounded, hands-on workshops for middle school students and expand their awareness of the wide array of college majors available at the university.

The urban agriculture program was designed for and conducted within PEO-PLE's high school scholars program, which is much different from PEOPLE's programs for younger students. It expands to include not only students in the Madison Metropolitan School District but also eligible students attending Wisconsin public schools with high numbers of low-income and minority students, specifically those in Milwaukee, Racine, Kenosha, and Waukesha, and all schools that serve federally recognized Native tribes. The high school program also focuses much more intensively on college preparation. It is structured around a four-year curriculum plan to enable "an increased understanding for college life and expectations as well as improved confidence in [students'] academic abilities and preparation."

During the first two summers of the high school scholars program, incoming sophomores attend a one-week program, and incoming juniors attend a three-week residential program, on the UW–Madison campus that includes "math, study skills, and writing skills development; ACT [American College Testing] preparation; workshops in the biological and physical sciences [where the urban agriculture track is housed], engineering, biomedical research and health sciences; and an evening curriculum in the fine and performing arts."

During the third summer, incoming high school seniors engage in a six-week residential internship and research experience for "learning and applying methods of scientific inquiry, analysis, and research in humanities and social sciences; hands-on experience and exposure to various professional fields through placements with hospitals, media companies, area businesses, and the UW." Finally, upon high school graduation and admission to UW–Madison, students participate in the eight-week bridge-to-college program, taking courses for college credit, and are given an experiential orientation to university life as an undergraduate.

In addition to PEOPLE's success rate, its formal connection with UW–Madison's College of Agricultural and Life Sciences piqued the CRFS staff's interest in a collaborative partnership. This college is fundamental to PEOPLE's third-year summer program because it provides internships for incoming seniors during the six-week summer program. However, despite the internship experiences that were available, the percentage of PEOPLE students declaring agricultural majors upon UW admittance remained low. This information was certainly not a surprise to the CRFS staff; the underlying foundations for the project's research included a knowledge of the historical trends that show how uninterested and uninvolved urban communities have been in the food system. Starting with the increased industrialization and mechanization of farming, which promoted migration from rural areas to cities, urban communities have become increasingly detached from agriculture. This detachment, the departure of infrastructure and community investment that resulted from white flight, and the disproportionate impact of economic downturns on communities of color have left inner-city communities especially vulnerable to issues surrounding food sovereignty and food access.

Considering these facts, in 2011 the CRFS project proposed to expand PEOPLE's science programs by developing an additional option based on urban agriculture. This new program was designed to provide students with relevant information and experiences in familiar landscapes and to impart valuable knowledge about this invigorated movement's role in community food systems. The abstract of the urban agriculture program states, "Students will learn how food systems work within the scope of five main topics. These topics are Production, Economics, Public Policy, Social Relations, and Food & Nutrition. The class will have an overarching theme that ties all of these main topics together in a comprehensive manner. Careers and job creation opportunities in the food systems industry will also be emphasized throughout the course. Hands-on activities and field trips to relevant sites such as urban gardens and dairy operations will be integrated with classroom learning."

The CRFS project also developed a "learning outcomes and activities guide" for each topic area that was inclusive yet thematic, and it enabled a revealing educational experience about where food comes from and how it gets from the field to students' homes.

The proposal was quickly accepted by PEOPLE's leadership, including former executive director Jacqueline DeWalt, who for years has been dedicated to expanding students' knowledge of and interest in possible college and career pathways. A pertinent problem that PEOPLE works to solve is the absence of early exposure; this exposure and related experiences are critical factors to increasing interest.

The majority of PEOPLE students come from urban backgrounds, so agriculture is often a mystifying topic for them. Lacking the knowledge of how food systems operate, urban students often equate agriculture exclusively with farming, a subject they find relatively foreign and usually unattractive. DeWalt and the CRFS project staff set out to construct a curriculum that provided adequate exposure and contained practical, culturally appropriate, relatable information that would help drive home the importance of urban agriculture and the food system as a whole.

The initial plan for integrating the urban agriculture program into all three years of the PEOPLE high school scholars program required the development of two separate curricula: a one-week science workshop in the first year (known as PEOPLE 1), and a three-week science workshop in the second year (known as PEOPLE 2). After two years of executing curricula for PEOPLE 1 and PEOPLE 2, the CRFS project also had to make accommodations for a six-week internship plan during the third year.

The Curriculum for PEOPLE 1

PEOPLE's urban agriculture program was launched during the summer of 2012. (In 2013 and 2014, one-week and three-week programs were offered; subsequently the program has been offered only during the three-week session, using similar content and concepts.) The program originally comprised a one-week workshop that encapsulated the primary goals of PEOPLE and the CRFS project: to provide students with a realistic way of exploring urban agriculture to increase their knowledge of additional career options while facilitating college readiness. The plan incorporated the key realms of food systems and urban agriculture listed in the abstract of the urban agriculture program.

FIGURE 22. PEOPLE students preparing a pizza. (Courtesy of University of Wisconsin–Madison PEOPLE)

By the end of the workshop, students were expected to be able to do the following:

- Develop learning experiences that create an understanding of urban agriculture and its components.
- Understand, apply, and personalize research on urban agriculture through reflection journals and a final project.
- Work collaboratively with other students, the instructor, and the community to increase their understanding of urban agriculture.
- Show an increased understanding of science and math concepts.

With this paradigm in mind, CRFS produced a curriculum with a theme of culinary production in an attempt to introduce students to thinking about food systems by focusing on a familiar product: pizza. Recognizing the four major components of a pizza—bread, sauce, cheese, and toppings—four days of the workshop were dedicated to the production processes, one day for each component (fig. 22).

On the first workshop day, before starting the pizza curriculum, PEOPLE 1 students toured Growing Power in Milwaukee to get a firsthand look at the daily activities of an established, functioning urban farm. From what teachers of the urban agriculture class gathered from students' journal entries, the students were unaware of terms such as *vertical farming* and *aquaponics*. One of the objectives was to debunk the students' preconceived notions of agriculture, and the Growing Power tour helped to do just that.

The second day, Bread Day, consisted of a food system modeling exercise that required students to pick a favorite food, list the ingredients in it, and then map out how far all the ingredients traveled from the farm to the processing facility, to the market, and finally home. To connect to this theme of food miles and the agricultural input required to create food products, the students then conducted a wheat conversion exercise that built on their mathematical skills. They determined the amount of wheat they would need to create specific pizza dimensions and calculated how much wheat would be required to feed a city the size of Madison or Milwaukee.

The third day, Sauce Day, consisted of an in-depth lecture on sustainability, food sovereignty, and the disproportionate food insecurity of certain communities. In keeping with the theme of the day, the students took a trip to the Goodman Community Center in Madison, where they participated in a scavenger hunt for pasta sauce ingredients in the center's garden plots, assisted by the center's youth volunteers and staff.

The fourth day, Cheese Day, included tours of the Dairy Cattle Center and Babcock Dairy Plant. On the latter tour the students were introduced to the various processes of milk and cheese production. After the tour a guest speaker from a local dairy discussed the food system from a business management perspective. Because the students had already been introduced to certain processing units and farms that are present in food systems, the day was organized so that the students could build on this knowledge. This enabled them to recognize the large corporations that are present in the food system as well as the roles of smaller, equally important local businesses.

The fifth day of the workshop was organized around community-supported agriculture and pizza toppings to drive home the various components of urban agriculture and community food systems. The day began with an overview of what community-supported agriculture is, its importance, and how it was relevant to the course and to community food systems as a whole. This topic provided an additional opportunity to discuss food sovereignty, augmenting the discussion from earlier in the course. The students also toured a local CSA operation.

The sixth and last day of the workshop challenged the students to use the knowledge they had acquired to produce an edible pizza and create a marketing scheme to accompany it. The students were prompted to think about cost efficiency when making their pizzas, which then pushed them to think about the costs related to harvesting, processing, and shipping that occur between farm and fork.

The Curriculum for PEOPLE 2

When constructing the curriculum for PEOPLE's three-week urban agriculture program, which began in the summer of 2013, the teachers recognized the difficulty of creating daily themes for a fifteen-day course. Consequently the teachers decided to create weekly objectives or themes that the curriculum would fulfill. The class instructors were supplied with a list of possible tours, speakers, and exercises appropriate for each week, but they were not limited to those choices. Rather, they were encouraged to develop their own curricula by drawing on their personal expertise to meet each week's objectives.

Similar to the first day of PEOPLE 1, the first week of PEOPLE 2 expanded on what urban agriculture is and what it looks like. We expected the students to gain insight into what current community food systems look like in Madison, including the identity and nature of the organizations within the systems and how those organizations collaborate and communicate with one another. Another objective for the week was to identify the scientific and business aspects of urban agriculture and the food system as a whole. Our hope was to expose students to possible careers in urban agriculture, whether in farming, running a CSA operation, working in food science, or managing a local restaurant.

A main goal of the second week was for students to receive skills in practical business management, marketing, and economics, with an emphasis on how these concepts are relevant in local and global agriculture. This instruction also provided an opportunity to integrate math lessons into the material.

Finally, the third week focused on the community aspects of agriculture. As in PEOPLE 1, the main goal of this week was an increased awareness of food justice and food sovereignty and how these concepts affect specific communities. The students were required to participate in exercises and discussions that demonstrated the importance of these concepts. This portion of the course was heavily focused on solutions to the issues discussed; the instructors highlighted how every lesson learned from the tours, exercises, and class speakers during the course contributed to these solutions.

Outcomes and Challenges

Overall, PEOPLE's urban agriculture program has been extremely successful, thanks in large part to the help of CRFS project staff members with expertise in curriculum development and planning. They executed a precise method for how to orchestrate lesson plans in a way that gave instructors the freedom to be creative while ensuring that the students were engaged. The urban agriculture program was implemented for five years during the CRFS project, and efforts are under way to make it a permanent part of the PEOPLE curriculum.

One challenge is that student responses to evaluation surveys were collected only sporadically, so we cannot provide complete empirical data about the program outcomes. However, the teachers reported informally that students who were initially uninterested in the program seemed to become more willingly engaged throughout the course—especially during PEOPLE 2.

After the second year of running PEOPLE 1 in 2013, an informal survey of the students in the course showed that they were dissatisfied with the number of field trips and tours that were crammed into one week—they felt "moved around too much." This was unexpected news that caused extra work in reorganizing the curriculum for the following year. However, we were excited that the students positively evaluated the in-class learning exercises. With this information in mind, the program consisted of more discussions and exercises in 2014. We learned that both students and instructors benefited from this change; students had more time to learn concepts, and instructors gained more freedom to create their own curricula. We have continued to keep these changes in mind as we plan the future of the program.

Future Opportunities for Growth and Expansion

In the last two years that the CRFS project supported the urban agriculture program, the PEOPLE staff, pleased with the program's growth, asked the CRFS project to develop an internship program for its incoming high school senior PEOPLE students. We launched a structured program for agricultural entrepreneur internships in collaboration with local food system organizations: the South Madison Farmers' Market in 2015 and Mentoring Positives in 2016. Student reactions have been uniformly positive, and the students' efforts may result in marketable products.

With the future in mind, the CRFS and PEOPLE staff are working together to better serve PEOPLE students through the urban agriculture program. Fund-raising is under way to support both the PEOPLE 2 curriculum and agricultural entrepreneur internships.

A major theme through this book is the idea that community engagement is a vital tool in establishing food security in urban areas. Urban communities must become deeply rooted and involved with both global and local food systems to ensure that they are not disproportionately excluded from receiving healthy, sustainable, and culturally appropriate food. Education is one approach to motivating individuals and organizations to engage in the food system. We hope that by increasing knowledge of how food production and distribution processes work and by illustrating how one can become further involved in them, we can ignite an interest in building food systems that serve communities, not corporate interests.

The authors thank Allison Dungan and George Reistad for their contributions to the development of PEOPLE's urban agriculture program.

CHAPTER TWELVE

Community and Regional Food Systems Policy and Planning

LINDSEY DAY-FARNSWORTH AND MARGARET KROME

In this chapter Lindsey Day-Farnsworth and Margaret Krome review the significance of federal food system policies, outline some of the ways that policies at all levels of government affect food systems, and discuss ways to change food policies, particularly through the actions of local food councils. This is another example of system thinking and collaboration that can positively affect all the CRFS values.

How can communities best act to promote community food security? This chapter strives to answer the question by exploring a variety of approaches to food system policy and planning. Wilde (2013, 1) defines food policy as "laws, regulations, decisions, and actions by governments and other institutions that affect food production, distribution, and consumption." We would extend this definition to also include access and disposal. Policy is also critical to the development of a food system infrastructure through a combination of incentives, direct investment, research funding, and training. Failure to fund public programs, or the omission of activities, can have policy implications just as specific actions do, according to food activist Mark Winne (Harper et al. 2009). In the United States, public food policy is created through legislation and regulation developed and administered across multiple jurisdictions and agencies. Understanding the distinctions in food policy at the federal, state, and local levels, as well as how food policy interacts across jurisdictions, can help inform strategic policy action at every level.

US food policy falls into a few broad categories: authorizing legislation, annual appropriations processes, and administrative implementation of the authorizations passed by Congress and signed into law by the president.

Authorizations

Authorizations are the congressional processes that establish federal programs, set the broad policies within which they work, and provide funding authority for the programs. Undertaken by the House and Senate committees with relevant jurisdiction, authorizations can be quite comprehensive and therefore are typically worked on every several years. For example, in agriculture and food systems, the House Agriculture Committee and the Senate Agriculture, Nutrition, and Forestry Committee have responsibility for the Farm Bill, which is reauthorized approximately every five years. Both committees have numerous subcommittees, which conduct hearings and develop the particulars of the bill's various areas, called *titles*. The 2014 Farm Bill had twelve titles: commodities; conservation; trade; nutrition; credit; rural development; research, extension, and related matters; forestry; energy; specialty crops and horticulture; crop insurance; and miscellaneous.

Other authorizations relevant to community and regional food systems are the Child Nutrition and WIC Reauthorization, passed approximately every five years, and the Food Safety Modernization Act, which was signed into law in 2011 as an update to previous food safety and cosmetic legislation but has no typical period of reauthorization. The Child Nutrition Reauthorization comprises policies authorizing the federal programs on child nutrition, such as School Breakfast, National School Lunch, Child and Adult Care Food, Summer Food Service, Fresh Fruit and Vegetable, and WIC. The Food Safety Modernization Act was a multiyear effort to respond to growing food contamination issues, and its provisions affect farmers, the food processing industry, aggregators such as food hubs, community-supported agriculture, and other parts of the food delivery system.

In addition to setting policy, authorizations set up the funding framework for the component programs through three main designations. Some programs' funding levels are directly mandated in the authorizing legislation; this *mandatory funding*, also called *direct funding*, is normally not subject to the discretion of the lawmakers in the annual appropriations process. Some major programs' mandatory funding is enshrined in *entitlement status*, in which a statutorily designated class of beneficiaries is guaranteed the benefits of that program. Entitlement examples in the Farm Bill include SNAP, commodity support payments, and crop insurance benefits. Because of the potential budgetary uncertainty imposed by such programs, there are few entitlement programs, but they can constitute an authorization's largest outlays. The third designation, *discretionary funding*, represents programs for which an authorization typically provides a ceiling of potential funding but whose actual funding level is determined annually by the appropriators; this is the smallest portion of outlays for most sectors of the federal government.

Appropriations

Annual appropriations are governed by the House and Senate Appropriations Committees and their subcommittees. The total amounts these committees can appropriate is determined each year by the Budget Committee of each body, before the appropriators begin making their decisions. The process is initiated each winter, usually in February, when the president issues the proposed budget for the next fiscal year that starts October 1; although the president's budget is nonbinding, it sets the terms of the debates that follow.

By law Congress must pass a budget for the upcoming fiscal year before the end of the current one, on September 30. In recent years this has rarely occurred, and instead Congress has usually addressed the problem temporarily by passing short-term legislation called a Continuing Resolution to fund the government on a status quo basis.

Although appropriations bills are supposed to focus only on the programs authorized as discretionary programs, sometimes Congress employs a mechanism called Changes in Mandatory Spending to cut programs mandated by one authorization or another. For example, this mechanism has been used to make big cuts to mandatory conservation program funding passed in the Farm Bill. Appropriations bills are also supposed to address funding only, not policy questions, but they have become increasingly subject to politicized efforts by constituencies that want to attach policy riders to appropriations measures, which encumbers the process.

Any federal legislation, whether an authorization, appropriations, or a stand-alone bill, follows a relatively predictable path. Once introduced in either the House or the Senate and referred to the committee with jurisdiction, the bill may be further referred to the relevant subcommittee. In the case of appropriations, much of the decision making occurs in subcommittees. After the committee marks up (i.e., votes on) the legislation or the components over which it has jurisdiction, the legislation goes to the full membership for a floor vote; at this time amendments may be offered, sometimes limited in number and kind by the parliamentarian. After both the House and the Senate have completed this process, their leaders choose conferees to negotiate any differences between the two versions. After the resolution of those differences, the final bill is voted on by both bodies and goes to the president to sign or veto.

Administrative Implementation

After an authorization bill has passed, the policy focus shifts from Congress to the agency or agencies that implement its provisions. That agency has the responsibility for negotiating the details of how a program will actually be run, and it does so through a rule-making process that allows for public input. Based on informal meetings and information gathering, the agency develops its plan for a rule and publishes that plan in the Federal Register. This notice of proposed rule making is usually called a *proposed rule*. It follows a specific format, including an explanation of the topic and the length of the period for public comment. After the agency considers the public comments, it issues a *final rule*, which becomes the operating policy for that program or topic.

Sometimes individuals or businesses believe that an agency's regulatory apparatus has been wrongly applied to them. Their recourse varies by agency; some agencies more clearly explain the routes for appeal than others. Sometimes appeals are informal and require only discussions with the agency staff, explanations of concerns, and the staff's exercise of discretion. Other appeals are more formal and involve administrative law judges, specifically designated appeals bodies, or other quasi-judicial appeal mechanisms. Sometimes appeals result in decisions in state or federal judicial courts, especially if the issues are complex and appear to revolve around agency interpretations of the underlying statute.

The states typically use a framework similar to the federal process described above, in which legislative bodies create policies and determine funding for programs and then governmental agencies implement the legislation.

The Difference between Legislation and Regulation

Though closely related, legislation and regulations are critically different in ways that have important implications for people working on food policy change. Legislation is created by Congress and the state legislatures to address particular social, economic, and environmental concerns. Legislation can include detailed information on how specific acts are to be implemented and enforced. However, more often than not, legislation details the intent and basic mechanisms of a policy, then authorizes the appropriate administrative or regulatory agency (e.g., the USDA) to address any gaps in the legislation through the development and enforcement of regulations. As such, regulations have the force of law but are often not expressly specified in legislation.

This approach to the development of legislation and regulation has several benefits. First, it frees legislatures from getting mired in the minutiae of policy areas about which they may have limited knowledge. Second, although regulations are typically developed by administrative agencies, federal administrative law requires a public comment process (described above) in which people who may be affected by particular regulations can provide feedback to the regulatory agency. Subsequently, the agencies are generally required to consider and release written responses to all comments. The intent of this process is to balance expertise and transparency by entrusting the initial development of regulations to agency experts while also creating a corrective function through public comments.

Knowing whether a particular policy issue is legislative or administrative can provide critical insight into where and how to advocate for policy change. For example, in Wisconsin, several recently proposed administrative changes would affect SNAP eligibility within the state. These proposals include adding a photograph of the primary cardholder to all SNAP EBT cards, restricting what foods Wisconsinites can purchase with SNAP benefits (i.e., food stamps), and requiring drug testing for qualifying, able-bodied adult SNAP recipients. Although these proposed changes would impose requirements that exceed those mandated at the federal level, administrative policies can also result from state agencies' decisions to opt out of certain federal accommodations (Nick Heckman, Madison and Dane County food security policy analyst, personal communication, 2016).

For example, states can appeal to the USDA to waive work requirements and time limits associated with SNAP benefits for able-bodied adults in communities and states where there is high unemployment. For years Wisconsin held such waivers. However, the current state administration has decided not to continue to seek these waivers, effectively changing policy through administrative action (Heckman 2016). As a result of this policy change, approximately two-thirds of the 22,500 Wisconsinites who became subject to the change between April and June 2015 were denied SNAP benefits within three months for failing to meet the work requirement (Lieb 2016). Research has found that many of the people who have been denied SNAP benefits as a result of the rollback of the work rule waiver struggle to find jobs because of their low educational attainment, physical and psychological limitations, or transportation constraints, such as lacking a driver's license. Although policy advocates could contact their elected leaders about SNAP eligibility rules, they could also appeal to the Wisconsin Department of Health Services directly because this policy change did not require legislative action.

HOW DOES FOOD POLICY VARY ACROSS JURISDICTIONAL LEVELS?

US policy is created through a system in which governing authority is constitutionally divided between the federal government and state governments, and state governments delegate some aspects of their authority to local governments. As such, all three levels have unique powers and limitations, with implications for food policy advocates and for the development of community and regional food systems. The following sections provide a synopsis of the key distinctions among federal, state, and local governments relative to their policy-making authority.

Federal Government

The US government's powers are constitutionally limited to the following six actions: declaring war, raising an army and a navy, coining and regulating currency, levying taxes, regulating interstate commerce, and establishing standards for weights and measures. According to Leib (2012b, 6), of these federal powers, the three of greatest importance to food and agricultural policy are "the authority to regulate interstate commerce, the taxing power, and the ability to attach conditions to federal funds." The federal government's authority to regulate interstate commerce is significant because it limits the states from implementing protectionist local or regional pricing schemes (McCabe 2011). Meanwhile, the other two powers enable the federal government to levy taxes and stipulate the terms under which federal funds may be allocated to states (Leib 2012b).

State Government

Any authority not constitutionally granted to the federal government or "necessary and proper" for executing its express powers falls within the jurisdiction of the states (US Const. art. I, § 8, cl. 5). This means that the states have considerable freedom to innovate with policy, making them "laboratories of democracy," as Justice Louis

Brandeis famously noted (*New State Ice Co. v. Liebmann*, 285 U.S. 262 (1932)). The powers conferred to the states are largely derived from police power, which gives them primary authority to engage in or delegate policy making on issues regarding "the health, safety, welfare, and morals" of their citizens.

It is not uncommon for state-level policy innovation to percolate up to the federal level. For example, as Leib (2012b) notes, California's leadership in developing menu-labeling requirements led to federal changes in menu labeling. This example is important because it underscores how federal preemption—invalidation of a state law that conflicts with federal law—can occur as policies that fall under federal jurisdiction supersede state and local policies once they are passed at the federal level. State and federal laws can establish limits on local government policies and on their scope or stringency. Knowing which state and federal powers particular policies may be subject to helps food policy advocates function better in the political landscape and anticipate where they may encounter preemption of local efforts.

In addition to creating policy, the states play an important role in the implementation of federal food policy, because many federally funded food and agricultural programs are operated at the state level or below. Despite some variation in name and function, five common state agencies are likely to serve as important partners in food and agricultural policy work: the departments of agriculture, public health, education, human services, and environmental protection (Leib 2012b). For example, state departments of agriculture often administer grant programs financed by USDA funds to support specific types of agricultural production, conservation, and food system development. In Wisconsin, the Specialty Crop Block Grant allocates funds to eligible producers and industry members through competitive grants intended to "enhance the competitiveness" of noncommodity crops such as fruits, vegetables, and tree nuts (Wisconsin Department of Agriculture, Trade, and Consumer Protection 2014, 3).

Meanwhile, state and local health and human service agencies oversee SNAP and WIC, which are funded through the Farm Bill, and school meals programs, which are financed through the federal Child Nutrition Act (Wilde 2013). In some instances, federal legislation requires specific types of matching funds by the states in order to receive federal support. Thus, the states can compromise individual and household access to federal programs by refusing to provide the required match. For example, the 2014 Farm Bill reauthorization required the states that participate in federal "heat-and-eat" programs to increase their per-household contribution for heating costs to $20; state contributions were previously as low as $1 per household. Faced with this additional expenditure, some states rallied to secure the funds necessary to maintain the program; other states, including Wisconsin, Michigan, and New Jersey, did not (Opoien 2014). The result was a decrease in SNAP benefits for low-income families in those states.

As this example illustrates, effective policy advocacy must occur at multiple levels. Overemphasis on federal-level legislation can distract from advocacy at the state and local levels. Conversely, focusing primarily on the state and local levels can result in missed opportunities to effect systemic change through federal legislation and budgetary processes.

Local Government

Local governments are creatures of the states and have no express power under the US Constitution. Unlike federalism, which confers different powers and responsibilities on federal and state government, *all* local government authority is ultimately granted and constrained by state government. Nevertheless, the delegation of power to local government varies considerably from state to state and across different types of local government (Leib 2012a).

The nature and types of authority held by a local government are influenced by whether it is located in a Dillon's Rule state or a Home Rule state (Leib 2012a). In Dillon's Rule states, local governments are limited to expressly granted powers, clearly implied powers, and powers necessary to perform specific local governmental functions. In contrast, Home Rule states confer much greater discretion on local governments regarding their system of governance and local policy making. However, this discretion can vary significantly for different types of local government.

For example, in Wisconsin, general-purpose local government includes counties, cities, villages, and towns, yet their degrees of home rule vary considerably (Anderson 2005). Cities have the highest degree of home rule, followed by villages. Counties and towns do not have home rule and are limited to the powers expressly granted to them by state statute. Understanding the local government structure and authority enables policy advocates to know what level of government to target, depending on the issue in question.

Local jurisdictions have land use zoning powers. Differences in their policy-making processes, authority, and interpretation of state laws such as zoning authorization are likely to affect the shape and scope of local food policy and planning. Consequently, food policy advocates might reasonably pursue zoning reforms to promote community and regional food systems development in multiple jurisdictions and at multiple levels of local government. Efforts to reform local laws to permit food production in developed areas, particularly for programs financed with local tax dollars, will probably fare better in communities with broader policy-making authority and tax-collecting powers. Nevertheless, just as the federal government has some powers to preempt state policy, Home Rule states have the authority to preempt local laws that conflict with state laws or stray into state jurisdiction.

In Wisconsin, as across the United States, the use of hoop houses (sometimes called *high tunnels*) has become a common means for urban farms to extend the growing season, especially in colder climates. The increased use of hoop houses, however, has not been accompanied by an increase in supportive municipal policy. Their lack of a fixed structure and foundation, the ability to dismantle and relocate them, and their independence from utility connections means that from a regulatory view, hoop houses fall ambiguously between a permanent structure and a utility shed. When hoop houses are considered occupied work spaces, regulators err on the side of safety and treat them as permanent structures covered by existing codes—that is, as enclosed spaces subject to fire, snow and wind loads, and other stresses. This treatment leads to inconsistent regulatory practices.

In 2013 the city of Madison incorporated new supporting language for urban agriculture in its rewritten zoning ordinance, permitting farming in different zoning districts. In permitting hoop houses, however, the facilitative zoning language in Madison is in conflict with state building codes. City building inspectors have interpreted hoop houses to be temporary structures based on fire-related concerns, and they will issue permits allowing them to stand for a maximum of 180 days in a 365-day period, obviating their season-extending possibilities. In Milwaukee, city officials—no doubt influenced by the presence of Growing Power and other local urban farms—have been more accommodating. Milwaukee views hoop houses as nonoccupied structures—essentially accessory structures under the city's recent Commercial Farming Enterprises land category. This regulatory acceptance of hoop houses is consistent with Milwaukee's generally advanced attitude toward urban agriculture.

Table 12 offers examples of different types of policy and regulation, spanning production to waste recovery, which can be used to support and promote healthy community and regional food systems. Depending on the nature of the policy or regulation and the funding sources, these activities may require action at multiple jurisdictional levels, including state and federal (Muller et al. 2009). The US Department of Housing and Urban Development's long-standing Community

TABLE 12. Local Policy Interventions to Promote Vibrant Community and Regional Food Systems

Phase of Food System	Land-Use Controls	Economic Development Incentives	Licensing and Regulation	Programs and Services
Production	Institute urban agriculture ordinances that designate appropriate types of agriculture as approved land uses.	Institute local food procurement policies for city departments.	Permit on-site produce sales at urban commercial or market gardens.	Provide vegetable gardening classes and resources through municipal parks and recreation departments.
Processing	Implement industrial retention through zoning and comprehensive planning to preserve sites for food manufacturing in metropolitan areas.	Establish agricultural processing renaissance zones. Offer incentives to certified commercial kitchens that make their space available to schools and food business entrepreneurs.	Publicize state-level cottage industry laws that permit limited sales of home-processed foods.	Provide cooking and food preservation classes through municipal parks and recreation departments.
Distribution and retailing	Make farmers' markets approved land uses and relax zoning codes to allow them to locate in a wider range of sites in residential zones and in commercial zones without public hearings. Provide flexible zoning regulations (e.g., set-backs, parking requirements,	Leverage municipal and/or county resources to fund or conduct feasibility studies of metro-area food hubs. Leverage CDBG funding to increase produce offerings at bodegas and corner stores in underserved neighborhoods.	Provide grocery store attraction incentives for underserved areas (e.g., fast-track permits for full-service grocery stores).	Promote the establishment of farmers' markets on city-owned land, including parks and public schools.

	and height restrictions) for grocery stores locating in underserved areas.			
Consumption	Provide long-term lease options for community gardens on public land.	Increase local food offerings by public school food service directors through the USDA's "geographic preference" option.	Ease or streamline licensing requirements for new farm stands, farmers' markets, and healthy food carts.	Provide electronic benefit transfer card machines to make farmers' markets accessible to WIC and SNAP recipients.
Resources and waste management	Define yard waste and food scrap composting as an agricultural, not industrial, land use to reduce the regulatory burden on urban agricultural composters. Align state and municipal composting land-use regulations to streamline the permit process.	Institute mandatory recycling and composting.	Differentiate regulatory and permit requirements for agricultural (e.g., yard waste and food scraps) and solid-waste composting operations.	Implement a municipal household composting program. Provide composting bins to residents and businesses.

Courtesy of www.extension.org/pages/70527/local-food-system-policy#.VbZA3oVp_Vs.

Development Block Grant (CDBG) program, for instance, has been creatively used by municipalities for food system support.

MECHANISMS FOR DEVELOPING COMMUNITY AND REGIONAL FOOD SYSTEMS POLICY

Community food security proponents note that community-driven problem-solving strategies can be more effective than top-down strategies for several reasons (Siedenburg and Pothukuchi 2002). First, they draw on local knowledge of needs and assets. Second, they have the potential to build capacity and community leadership. Third, they cultivate a greater sense of community ownership over a given solution. Consequently, meaningful grassroots involvement is critical to developing long-term strategies for addressing complex food system issues.

Given the policy context outlined above, how can residents and community organizations collaborate with other institutions to promote policies that will foster more economically vibrant, equitable, and sustainable food systems? The remainder of this chapter highlights various approaches, including community and regional planning, food policy councils, food policy audits, and strategies for connecting the grassroots with federal-level policy work.

Community and Regional Food Systems Planning

Community and regional food systems planning is "the collaborative planning process of developing and implementing local and regional land use, economic development, public health, and environmental goals, programs, and policies . . . to enhance the overall public, social, ecological, and economic health of communities" (American Planning Association n.d.).

Since 2000, when the professional planning field began to reengage with community and regional food systems (Pothukuchi and Kaufman 2000), planners have tackled a wide range of activities, including the following:

- Defining food systems issues for planning, policy, and implementation (American Planning Association 2007)
- Using regulatory tools to promote urban agriculture, mobile vending, and healthy food environments (Mukherji and Morales 2010; Morales and Kettles 2009a, 2009b; Hodgson, Caton Campbell, and Bailkey 2011; and Neuner, Kelly, and Raja 2011)

- Mapping and geographic analysis of the food system (Block and Kouba 2006; Peters et al. 2009, 2012; Raja, Ma, and Yadav 2008; Meter 2009)
- Promoting multistakeholder organizations and processes to improve problem solving (Pothukuchi and Kaufman 1999; Barling, Lang, and Caraher 2002; Koc et al. 2008; Muller et al. 2009).
- Incorporating agrifood components into planning documents (Raja, Born, and Russell 2008; Neuner Kelly, and Raja. 2011)
- Conducting research to support development of food system infrastructure (Skeo Solutions 2012; Cantrell et al. 2013)
- Supporting programs and community-driven projects (Caton Campbell and Salus 2003; Cobb and Houston 2011)

Despite the growing involvement of municipal planners in community and regional food systems, food system planning continues to be a professionally diverse field; many of the people engaged in it work in other local government agencies (e.g., public health) or for nonprofit organizations or Cooperative Extension Service offices. This hybrid nature of food system planning creates both challenges and opportunities. Grassroots-based initiatives present opportunities for community leadership and priority setting. Yet without strategic power analysis or access to local decision makers, it can be difficult to realize community-driven planning efforts because of limited resources and political authority. Conversely, local governments may have planning authority and access to resources, but without strong ties to the community they risk irrelevance and miss opportunities to foster civic engagement. Professionalized food system planning is well documented in peer-reviewed and "gray" literature (public agency and NGO-produced documents that have not undergone peer review); the first section highlights an example of neighborhood-scale food system planning based on a community-organizing model.

Dudley Grows: Community-Based Neighborhood Food System Planning at a Glance

LINDSEY DAY-FARNSWORTH

Dudley Grows, as its literature describes, "envisions a local resident-led food system that provides access to nutritious, affordable healthy food to all our neighbors,

FIGURE 23. Neighborhood grower Sayed Mohamed-Nour tending to communal plots in the Dudley Greenhouse, Boston. (Courtesy of Travis Watson)

brings economic opportunities to residents, and protects the environment." This initiative is a collaboration of three organizations with long histories of youth development and organizing in Boston's Dudley neighborhood: the Dudley Street Neighborhood Initiative (DSNI), Alternatives for Community and Environment (ACE), and The Food Project.

The Dudley neighborhood is racially and ethnically diverse: 31 percent of the population is black or African American, 28 percent is Latino, 28 percent is Cape Verdean, and 14 percent is white. Also, 41 percent of its households identify their primary language as other than English, and 34 percent live below the federal poverty level. Despite its economic challenges, this neighborhood has received national attention for its effective community organizing around vacant land, housing, and environmental and food justice and its high levels of resident engagement in neighborhood planning and development (fig. 23).

From its start in 2014 as a neighborhood food system planning initiative, Dudley Grows has built on momentum from previous local projects. Its objective is to give neighborhood residents an opportunity to articulate a vision for the neighborhood

FIGURE 24. Dudley Grows steering committee member Ivelise Rivera facilitating a conversation with other residents. (Courtesy of Travis Watson)

food system and to incorporate existing food and agricultural activities into a cohesive plan. Staff members from the three organizations shepherded the planning process with guidance from a steering committee of residents and local businesses and implementation support from youth employees (fig. 24).

The planning consisted of three components: (1) youth-led surveys of residents, community market owners, and gardeners; (2) community engagement events ranging from listening sessions to neighborhood celebrations; and (3) monthly steering committee meetings to reflect on the findings and discuss the next steps. By the end of the nine-month process, Dudley Grows had identified the following five priority areas within an overarching theme of local job and income creation:

1. Build a resident-owned supply chain for great food in the neighborhood, which would help grow food businesses that create neighborhood wealth and jobs.

2. Permanently secure vacant land for gardening by interested residents, so

that those who want to produce food for themselves or the neighborhood can do so.

3. Improve the food in neighborhood schools, ensuring that youngsters who eat at school are well nourished with food they enjoy.

4. Expand access to great food for lower-income residents, building creative new ways to make great food affordable to all.

5. Encourage physical development to support the neighborhood food system, and advocate for food interests in planning, building, and community development.

Through a CRFS Innovation Fund grant to The Food Project, the Dudley Grows initiative supported the development of a resident-owned retail and wholesale produce business. In the spring of 2015 The Food Project explored product preferences and price points with neighborhood retailers and developed a short list of pilot products: collards, corn, cabbage, cilantro, zucchini, scallions, and green peppers. That fall The Food Project helped the wholesaler acquire a walk-in cooler and began investigating funding to finance facade improvement and a SNAP-matching program for the market. Plans for the 2016 season included continued preseason planning and a youth-led produce marketing program.

The Food Project and DSNI convened another neighborhood steering committee to guide implementation. A youth-led door-knocking campaign gauged the residents' level of interest in converting vacant lots into small-scale market gardens: gardens for local retail and wholesale markets. DSNI has eminent domain over a portion of the Dudley neighborhood, which gives it unique power over land use decisions for vacant lots. When asked how this neighborhood planning process differed from city-led planning, The Food Project's Boston regional director, Sutton Kiplinger, and DSNI community organizer Bayoán Rossello-Cornier responded that there is better turnout to planning-related events, a higher level of trust between project leaders and community residents, closer communication between residents and support organizations, and more sustained involvement of residents and organizational partners.

"DSNI and ACE are organizing operations," Kiplinger explained. "They've got a good turnout game—if the city wanted to organize a process in this neighborhood, [it] would call DSNI and ask [it] to get people there."

The city actually does this "all the time," Rossello-Cornier added. "There's a level of comfort with [the neighborhood] people because of the trust that exists between us [DSNI] and them, which does not exist between them and the city." This trust is rooted in the role that DSNI has historically served in the neighborhood. For example, Rossello-Cornier noted that before DSNI starts a project, it determines whether the project is motivated by community interests, personal interests, or outsiders. If an issue is not motivated by the community, DSNI doesn't pursue it. Correspondingly, resident involvement is not about a onetime community listening session, but, as Kiplinger described, "it's about sustained resident engagement, so even if we had people who weren't coming to all of the [neighborhood food system planning] meetings, they were very much in conversation with us through the whole thing, and they now are invested in . . . participating and making things happen."

In reflecting on lessons learned from the community-driven planning, Kiplinger remarked, "For me, it underscored how communities know what's good for them. And if they can have control of the neighborhood food system and the resources that it takes to build what they envision, what they envision is *exactly* what the public health community wants for them, but their articulation of it is so much more powerful than anything the public health community would be able to come up with. . . . I think that every community we look at as having challenges around its food system probably has a vision that would solve those problems, and it's just a matter of asking and then being serious about making sure that [the community] has the resources and the capacity to make its own decisions."

Food Policy Councils

Food policy councils are another strategy that community groups and local governments can employ to advance policy change at state and local levels. The first food policy council was created in 1982, but not until the rise and confluence of food movements around 2000 did food policy councils become widely recognized in planning and policy spheres. Sauer (2012) identified nearly two hundred state and local food policy councils in the United States.

There are several motivations for these coalitions. First, local and state governments administer food policy and programming from multiple agencies, which can result in policy misalignment and missed opportunities (Pothukuchi and Kaufman 1999). Food policy councils can identify ways to improve alignment and generate coordinated approaches to complex problems. Second, food policy councils have been identified as vehicles for promoting food democracy by opening up the problem-solving and policy-making process not only to food system practitioners but to engaged gardeners, eaters, and community activists (Winne 2008). Consequently, food policy councils often include members who represent a range of vantage points, including farmers, antihunger advocates, institutional food service operators, urban gardeners, public health experts, small food business operators, policy makers, food justice advocates, and active community members.

There is considerable variation in the structure and operation of food policy councils. They can be initiated by local or state governments or by grassroots groups, and they have been structured in a number of ways, ranging from appointment-based city committees to open-membership citizen coalitions to hybrid organizations. Nevertheless, most food policy councils convene diverse food system stakeholders to provide one or more of the following functions: "serve as forums for discussing food issues, foster coordination between sectors in the food system, evaluate and influence policy, and launch or support programs and services that address local needs" (Harper et al. 2009, 2).

A number of studies have examined the successes and challenges of food policy councils (Dahlberg et al. 1997; Borron 2003; Harper et al. 2009). Day-Farnsworth (2016) found that successful food policy councils leverage the expertise and networks of their professionally diverse members to effect meaningful policy change and program development by accelerating existing projects, incubating and vetting nascent projects, and fostering systems knowledge and partnerships. In particular, she highlighted the importance of food policy councils differentiating themselves from other organizations by exploring specific ways they can add value to existing activities by serving a connective and facilitative role.

Harper et al. (2009) provided an excellent summary of the major challenges facing food policy councils, which include working with complex political landscapes, evaluating the councils' impact, and dealing with a lack of stable funding and staff capacity. Packer (2014, 10) also found that some food policy councils have fallen short of their promise of inclusivity and civic leadership. Specifically, she observed that "professionalized and foundation-dependent" food policy councils "circumscribe an alienating form of 'participation' that appeals mainly to members who share the same class and educational backgrounds as their founders (and funders)"; this results in the de facto exclusion of already marginalized groups.

Despite these challenges, the evidence suggests that when food policy councils are adequately staffed and resourced, they can spur program development and policy change that reflects the strengths of their multiple stakeholders. The story of the Good Food Purchasing Program (see section) provides a compelling example of an integrated policy and program initiated by the Los Angeles Food Policy Council.

The Good Food Purchasing Program: Institutionalizing Good Food Values through Policy, Partnerships, and Supply Chain Innovation

COLLEEN MCKINNEY

The Good Food Purchasing Program (GFPP) harnesses the purchasing power of major institutions to encourage greater production of sustainably produced food, healthy eating, respect for workers' rights, humane treatment of animals, and support for the local small-business economy. It started in Los Angeles and in 2015 became a nationwide program.

The GFPP was developed by a Los Angeles Food Policy Council working group composed of local representatives and reviewed by prominent national organizations representing a diversity of food system issues and stakeholders. The equity-based framework this group developed is unique because it gives equal weight to five values: local economies, environmental sustainability, a valued workforce, animal welfare, and nutrition. The GFPP's tiered structure allows institutions the flexibility to participate anywhere from the baseline one-star level up to the high-bar five-star level, depending on goals and budget. This means that any institution can commit to using its purchasing power to have an impact, even if it has to take small steps at first.

The City of Los Angeles was the first adopter, committing to the GFPP through a mayoral executive directive and a City Council motion on Food Day 2012. The Los Angeles Unified School District (LAUSD), the second-largest purchaser of food in California, followed a few weeks later. Together these early adopters represent more than 750,000 daily meals.

The LAUSD quickly demonstrated its ability to influence its supply chain. Its produce and bread distributor, a company that provides more than three million school meals per day throughout the western United States, made immediate shifts to help the district meet its GFPP commitment. For example, the LAUSD redirected its produce dollars toward local producers, a shift that led to $12 million being spent in Southern California instead of outside the region and that created an estimated 150 new jobs. The distributor also brokered a relationship to work with Food Alliance–certified sustainable wheat farmers in California to become the primary source of grain for baking products for the school district, replacing wheat previously grown out of state and creating a more environmentally friendly, healthy product for students.

Over time this LAUSD distributor has continued to build infrastructure that supports its capacity to provide more local sustainable products for the district by transforming its internal tracking system of suppliers.

NATIONAL EXPANSION

The experience in Los Angeles demonstrated that the GFPP allows institutions to make more informed decisions about the suppliers they work with, increases the public's capacity to engage elected officials and ensure that taxpayer-funded food contracts reflect community values, and provides high-quality food to communities that need it most. Food policy councils, government administrators, and other local stakeholders across the country reached out to the Los Angeles Food Policy Council to learn from the experiences of the staff members and discuss what it would take to replicate the GFPP in their locations.

Interest in support from other locations and the replicability of the GFPP model prompted the council to spin the GFPP off into a national program. The Center for Good Food Purchasing was created to provide adoption and implementation support in cities across the United States, such as Oakland, Austin, Chicago, and San Francisco (where it was adopted by the San Francisco Unified School District in May 2016). Under this model the considerable infrastructure developed to implement the GFPP in Los Angeles can be replicated in other places while still reflecting local priorities and meeting local needs.

The potential for the Center for Good Food Purchasing initiative is enormous. Imagine the impact that dozens of major institutions could have if they demanded better products from their suppliers. Their collective impact could redirect billions of dollars to sustainable, fair suppliers and have ripple effects that could transform our food system. Good Food Purchasing can be a pathway for building sustainable and socially just regional food systems that revitalize local economies and enable residents to prosper.

Food Policy Audits

In the past fifteen years a variety of food assessments have been developed to identify needs and assets and to bring focus to planning and policy efforts among community and regional food systems. In a review of food system assessments, Freedgood, Pierce-Quiñonez, and Meter (2011) found eight distinct types: local and regional foodsheds, the comprehensive food system, community food security, community food asset mapping, food deserts, land inventory, the local food economy, and the food industry. However, O'Brien and Cobb (2012) rightly noted that none of these assessment types explicitly identify a spectrum of extant and potential local policy options. As this chapter has highlighted, understanding the policy environment is crucial to effecting change in the food system because policies and regulations have a supportive or restrictive impact on virtually any food system activity, from urban agriculture to donating food to composting. Recognizing this gap in the community food assessment literature, the University of Virginia Department of Urban and Environmental Planning faculty developed and piloted a new form of assessment: the food policy audit.

The intent of the food policy audit is to bring a planning lens to the food assessment model in order to identify gaps and contradictions in local food policy frameworks and to guide future policy and planning priorities. Building on the well-defined structure of the energy audit, the University of Virginia team developed 101 yes-or-no questions, which were organized into five topical sections: public health, economic development, environmental impact, social equity, and land conservation (O'Brien and Cobb 2012). The range of policy categories provided topical breadth, and the yes-or-no question structure reduced the ambiguity of the results. Planning students were tasked with reviewing the local planning and policy documents to identify whether and where particular types of policies existed. The first phase of the audit provided baseline data that the students then shared with community members in a second phase to validate and contextualize the results.

This flexible tool has been employed and adapted to other contexts. For example, Jill Clark (2015, 1) of Ohio State University sought to transform it into a "citizen-oriented assessment tool" by conducting a food policy audit in conjunction with the citizen-based Franklin County Local Food Council in central Ohio. Although food policy councils have gained currency as a civic mechanism for promoting food system change, many food policy councils and coalitions have struggled to find their footing in an already crowded organizational landscape. Clark's innovative use of the food policy audit strengthened the link between civic engagement and the development of tangible policy objectives by employing the audit as a tool to translate the mission of the Franklin County Local Food Council into a technically informed political agenda.

In addition, Clark (2015, 1) and her community colleagues improved upon the University of Virginia audit framework by quantifying the score, so that each yes response equaled one point. This modification enabled the audit tool to "serve not only as an agenda-setting tool for local food policy, but also as a quantifiable benchmarking tool for progress toward local food policy goals." As recent innovation shows, food policy audits can serve as a tool to promote civic engagement and to develop an evidence-based local food policy agenda. A food policy audit of the Milwaukee metropolitan area was supported by the CRFS project (see section).

Milwaukee's Local Food Policy Audit: Adapting Urban and Regional Audit Models to Promote Food Equity at the Metro Regional Scale

MARCIA CATON CAMPBELL

Milwaukee is renowned for fostering urban agriculture. It is home to many urban agriculture nonprofit organizations, most notably Growing Power, and 85 percent of the city is zoned for urban agriculture. The City of Milwaukee's *Citywide Policy Plan* (2010) makes many specific references to urban agriculture, community gardening, and farmers' markets. However, it discusses other food system issues, such as healthy food access, in only the broadest terms. In contrast, a number of other North American cities—Baltimore, Minneapolis, Cleveland, Portland (Oregon), Toronto, and Vancouver—have worked intensively on their food planning and policy environments.

In this respect Milwaukee lags behind other cities in assessing the food policy environment and its effects, positive or negative, on residents' health and well-being. Healthy food access is best understood as a system problem because the availability of fresh food is a function of many interconnected policies and activities throughout the food system (Hodgson 2012, 18). Consequently, a comprehensive, systems approach to food access can "shed light on how all sectors of a community's food system—production, processing, distribution, consumption, and water recovery—as well as a community's political, social, and economic environment, may be contributing to food security and food access issues" (see also Raja, Born, and Russell 2008).

The city's 2013 sustainability plan, ReFresh Milwaukee, acknowledges this gap in the city's food policy and planning and identifies the need for "a citywide food system policy and action agenda" (City of Milwaukee Environmental Collaboration Office 2013). However, the city's ability to guide government action and policy change to achieve these targets will require a more complete understanding of the food policy environment as well as targeted public conversations to identify priorities. The local food policy audit, conducted by the Center for Resilient Cities through a CRFS Innovation Fund grant, was designed to establish a policy baseline from which the city staff could propose policy changes to enhance access to healthy food for all Milwaukeeans.

The center's staff assembled the audit tool for Milwaukee from two versions of an open source audit tool: (1) the original tool developed and piloted in 2010 by the University of Virginia's Department of Urban and Environmental Planning (O'Brien and Cobb 2012); and (2) a variation adapted by the Mid-Ohio Regional Planning Commission and implemented by the Franklin County Local Food Council (2014).

The Milwaukee audit tool consists of 129 questions divided into four categories—equitable food access, land use and zoning, economic development, and public health—presented in spreadsheet form. The tool only reveals Milwaukee's existing food system policy infrastructure; it does not evaluate whether the policies that have been implemented have achieved a desired effect. To accompany the Milwaukee audit, the Center for Resilient Cities partnered with a spring 2015 graduate course in the Nelson Institute for Environmental Studies at the University of Wisconsin–Madison. The twelve students examined the food policy environment and infrastructure of select jurisdictions around Milwaukee, including the city of Racine, the city and county of Waukesha, and Ozaukee County. Their work helped situate the Milwaukee audit findings in a regional context.

The Milwaukee audit included the following major findings:

1. Equity and inclusion in Milwaukee's food policy should facilitate a city-level effort at articulating values, such as racial equity, that develop an evidence base related to healthy food access.

2. Food access can be solved only with a systemic approach; a systems understanding of healthy food access must be taken.

3. Effective strategies are based on good evidence. City staff and nonprofit partners should conduct a comprehensive, citywide food assessment to study barriers to healthy food access for each of Milwaukee's thirteen neighborhood strategic planning areas.

4. Infrastructure is a missing link in the development of commercial farming enterprises and in local food procurement. The city should take an inventory of the food aggregation, processing, storage, and distribution infrastructure.

5. Healthy food is an economic driver. The city should incentivize the sale and purchase of healthy foods through programs such as a healthy food retail initiative and ensure that other land use and economic policies support this aim. The city health department should articulate goals, benchmarks, and tracking measures for healthy food access and public health.

6. The city should encourage large institutions to increase their good food offerings and partner with nonprofit groups to engage hospitals, public school districts, colleges, and universities in a conversation that leads to the adoption of a good food purchasing policy. (See section on the Good Food Purchasing Program.)

7. Multimodal transportation is essential to healthy food access. The city, partnering with nonprofit groups, should map bicycle paths, pedestrian paths, and bus routes to improve access to healthy food.

8. The city should expand the pilot food waste collection program citywide and allow and regulate multiple scales of composting, with emphasis given to composting activities that fall between those regulated by the city (twenty-five to fifty cubic yards) and those regulated by the state (more than fifty cubic yards).

Much of this chapter focuses on policy activities at either the federal or the local level. Building a linkage between these two spheres can create meaningful mechanisms through which practitioners, community leaders, and residents can influence federal policy. This enables local observations and experimentation to inform changes to the regulations and programs that shape community and regional food systems development.

For example, in recent years policy coalitions such as the National Sustainable Agriculture Coalition (NSAC) have influenced policy in many areas, from Food Safety Modernization Act rules to Farm Bill policies. NSAC's network of farmers, food entrepreneurs, conservationists, and community groups translates firsthand experiences into federal programs that support work as diverse as farmland conservation and direct marketing (see section).

From the Grassroots to the Farm Bill: The National Sustainable Agriculture Coalition's Process for Gathering Local-Level Input

BY MARGARET KROME

An example of how individual citizens engage in federal policy is NSAC's typical approach to Farm Bill policy development and action.

Starting two or three years ahead of the probable passage of a farm bill, NSAC encourages its member groups to hold listening sessions with various constituencies on issues affected by this comprehensive legislation. The feedback that the member groups gain from their listening sessions is shared with NSAC's issue committees, which at the end of 2015 focused on the following: (1) research, education, and extension; (2) marketing, food systems, and rural development; (3) conservation, energy, and the environment; (4) farming opportunities and fair competition; and (5) food system integrity. NSAC also has a committee dedicated to issues of diversity and social justice. Each committee encourages its members to conduct listening sessions with diverse stakeholder audiences to identify concerns and priorities for action in the upcoming farm bill, in addition to reviewing research and data on the usage and effectiveness of a range of programs.

As issues and ideas accumulate, NSAC's committees begin to sort through them, finding synergies and setting priorities. Meanwhile, NSAC begins discussing these priorities with its partner organizations and coalitions. Together they identify allies on issues of common interest and engage diverse grassroots voices to advocate for policies that will advance sustainable agriculture and healthy community food systems.

NSAC and its partners then cluster the policy agendas into packages to discuss with members of Congress, preferably from both parties. Once there are cosponsors from both parties, one cosponsor may introduce a marker bill as a way to bring the issues it addresses into public discussion as the bill's consideration becomes more focused. Grassroots supporters begin encouraging members, especially those on the Agriculture Committees, to support these marker bills.

Both House and Senate Agriculture Committees and sometimes their subcommittees hold hearings on topics over which they have jurisdiction. NSAC may work with committee staff members to arrange for a grassroots spokesperson to testify on a topic. As the committees and their subcommittees begin serious consideration of Farm Bill proposals, grassroots members, especially those in states represented by Agriculture Committee members, send letters and call the offices of these members about proposals of importance to them. NSAC organizes grassroots "fly-ins," in which farmers and other stakeholders travel to Washington, DC, for direct meetings with their representatives or senators on the Agriculture Committee to discuss specific Farm Bill proposals. Other grassroots actions include using traditional media and social media to heighten the awareness of issues, specific proposals, and legislative actions as the Farm Bill moves through committees to each body's floor and then into conference negotiations on the final bill.

CONCLUSION

In the United States, food policy creation and implementation is a complex, participatory, and multilevel undertaking. Federal and state food policy processes determine the laws, regulatory frameworks, and funding allocations that shape our agricultural and nutritional activities. Knowledge of policy-making processes and the powers of different jurisdictions enables community food advocates to make informed decisions about where to intervene to influence policy on particular food issues. This chapter highlighted several specific ways that community food advocates can influence policy, including participation in formal public comment processes to recommend changes to regulations and engaging with policy coalitions such as NSAC to inform the Farm Bill and other legislation.

Community food advocates can also take numerous actions at the local and regional levels. For example, strategic assessment tools like policy audits and advisory bodies such as food policy councils can influence land use, transportation, and management decisions that affect agricultural production, access to healthy food, and waste recovery. Through a mix of local policy and organizing, community food advocates can also transform food procurement practices and supply chain development to better reflect values such as community economic development, humane animal treatment, and fair wages and safe labor conditions for food chain workers (see sections on the Good Food Purchasing Program and Dudley Grows).

Many food issues are system problems rather than technical or sectoral problems, so they require strategies that comprehend how different jurisdictions, policies, and strategies interact. Grassroots initiatives present opportunities for place-based problem solving, innovation, and community leadership. Involvement in federal and state policy making and administrative implementation provides opportunities to influence policy and regulatory frameworks. An overemphasis on federal-level legislation can distract from advocacy at the state or local level, whereas focusing solely on the local level can result in missed opportunities to influence federal resource allocation and rule making. By using and bridging a wide array of policy and planning tools, we can make the most of our toolbox and effect meaningful change from the grassroots and the grasstops.

CHAPTER THIRTEEN

Cultural Dissonance

Reframing Institutional Power

ERIKA ALLEN, RODGER COOLEY, AND LAURELL SIMS

In this chapter three authors working through the Chicago Food Policy Action Council describe their perspectives on institutional racism in food systems, including institutions of higher learning. They present ideas and principles for how community organizations and institutions can more effectively collaborate to address issues of food justice.

Applying the principles of social justice, equity, and antiracism can critically advance community food systems work. The prevailing voices in the sustainable food system movement tend to be white (Eurocentric), with people of color relegated to roles in food justice or food desert discussions by the dominant voices in our movement. This critical perspective is reflected in the current social change dialogues that have reached popular cultural spaces as well. Who tells the story is important. The reality is that we are fighting not only for a change in the quality of food but also for *all* the tenets of good food, which change depending on the cultural framing and narrative.

NARRATIVE OWNERSHIP AS A POWER-SHIFTING STRATEGY FOR COMMUNITY-BASED FOOD SYSTEM PRACTITIONERS

The discussion of good food is typically about health or taste and occasionally about farmers and issues of fair labor practices. In the definition of food justice it is important to include the context of labor, environmental impact, and policy issues that are connected to bringing good food to our tables, even though these tend to be separated and researched as singular issues. However, integrating social justice and altering the structural apparatus that makes it possible—for example, creating the ability to access organic grapes in a geographically defined food desert—will never change racism or the social and economic disparities based on race. Food justice may look to some like making sure that everyone has equal access to grapes. Many of us instead consider food justice to also include the ability to earn a living wage growing and harvesting those grapes, on land that is owned by the farmer, individually or cooperatively, and in an environment not compromised by chemicals. This kind of food justice cannot be achieved without changing the power dynamics that structural racism perpetuates within the agricultural industry and our culture as a whole.

What does this perspective mean in the context of research and academic thought? Even though some academic areas such as grounded theory and participatory action research are designed to learn from practice, more effort could be made to integrate antiracist theory within the fabric of inquiry. The narratives and rigor that researchers and academics generally use can distill the complexity of key issues into formats that are not helpful to people on the front lines of community work; without a framework for understanding the dynamics, biases, and perspectives, the white privilege lens is not adapted to effective interaction and communication. Without a significant effort to shift perspective, otherwise well-intended work can lead to the maintenance of structural racism and oppression, and stand in opposition to community efforts of finding, defining, and owning a food system that serves the needs of oppressed communities.

To put this in perspective, I (Erika) will describe my own experiences with Growing Power. Growing Power's approach to food access is through the development of community food systems that are owned by, operated by, and reflective of the cultures and cuisines of the people who live within those communities. This approach connects and exposes the diversity of food that can be grown to each generation we work with.

A community food system includes the entire length of the food chain, from producer to consumer. It encompasses every step of the growing, marketing, and distribution process, and it evaluates the inputs required and the outputs produced during all stages of the system. It ensures that the community's needs are met in a sustainable and acceptable fashion. The contemporary large-scale industrialized food system largely disregards the aspect of community and functions instead as a global enterprise requiring the unnecessary expenditure of many finite resources.

We are often reminded that Growing Power is one of the few community food system organizations led by people of color; it maintains a diverse staff and employs grassroots and community residents (those most affected by historic inequity) on a larger scale than most groups. It is a challenging mission to develop staff capacity across divergent backgrounds and to have them work together to build a new food system reality in spite of limited resources, internalized oppression, and cyclical poverty. We do this while simultaneously educating and forming strategic partnerships with top-down institutions such as funders, corporate entities, and universities that have not addressed their inherent structural racism to the degree necessary to form equitable partnerships.

The process of change and understanding is a long one. The need to traverse the cultural learning curve of power-wielding entities is often time-consuming and ineffective, since it requires the capacity to be present and have enough autonomy not to be shut out of developing economic opportunities, informing policies, or setting the direction and content of research and scholarship.

Even as an organization with access to resources and political goodwill, Growing Power must balance the day-to-day urgency inherent in food production and distribution with the need to function in academic and political spaces. It is a challenge to maintain voice and presence within a culture that is not hands-on by design.

Furthermore, the typical academic narrative and design of effecting change is not responsive but extractive—in other words, the process extracts data, story, and technique and reframes it as valid through institutional incorporation or codification. If we are to change the narrative, this approach should be reversed so that the community narrative and process are valued and considered to be on at least equal footing in its own perspective.

So let's begin to have more inquiries. For example, who is a community partner? What do we mean when we say *community*? Do we value and prioritize the knowledge, expertise, and voices of noninstitutional partners?

In some circles, *community* is a code word for people outside the dominant institution that controls the resources, which in this context could be money, research, information, technical assistance, or simply access. Institutions such as universities, the judicial system, and corporate entities often seek to "partner" with community entities—either nonprofit organizations or individuals active in the organizing of the communities that require services or assistance.

There is a prevailing belief within grassroots intellectual spaces that such big institutions cannot have equitable relationships with community-based organizations, especially those led by people of color. This belief is based on the manner in which top-down institutions and their actors are designed to uphold their superior position. Individuals are rewarded for participation and upholding their institutions and are censured and removed if they do not play by the rules. As a result of this codification of the institutional structure, the system as a whole is hard to change. There is often an acknowledgment of these issues, which may be addressed with a training event, a longer course of study, or the occasional development of a center focused on diversity, race, and class for discourse and study of the issues. There is, however, little power to actually shift the dynamics of white supremacy within the institution itself. In fact, the belief that its "expert" culture is addressing institutional racism perpetuates the status quo through self-validation.

Challenging the dynamics and culture of institutions that have codified white supremacy is dangerous for the agents within those institutions, regardless of their skin color or cultural background. People who raise challenges risk the loss of tenure eligibility and research funding, the ability to progress within the institution, and even continued employment. Without a systemic policy shift by the institution's executive leadership or governing board that commits to a course of understanding in how institutionalized racism affects and drives the way we learn, teach, and reach out to communities, we will continue to perpetuate systems of oppression and power that are protected and supported by our participation and that hinder the development of long-term partnerships based on real equity.

This is a conundrum. How do we transform our institutions to be able to partner with community organizations and leaders that share an understanding of these issues? We must develop new tools and a shared language to restructure the protocols, processes, and tools that measure success from the community's perspective but that do not perpetuate and regenerate the superior position of the institution.

A first transformative step is a basic set of training sessions dedicated to dismantling racism, an ongoing study of systems of oppression from a historical perspective, and a thoughtful, deep analysis of current popular thinking. These sessions should be offered to all who are engaged and invested in equitable partnerships with communities. A second step is to actually show up for the first step—not just on a literal level but also with the sincere intent to engage, participate, and apply the concepts, harnessing the power of privilege to undo and challenge racism's many manifestations. Often only a small percentage of an institution's employees get involved in this training because it is not prioritized. The partial attendance and lack of follow-up can mean that antiracist principles are never applied within the community partnership.

Community partners should certainly be able to expect that both overt and discreet racist practices will be eliminated or at least challenged as they arise with institutional colleagues. A simple example would be an institution's failure to provide a community with bilingual training materials and translators when it is common knowledge that English is not the community's primary language. Such a failure signifies that maintaining the institution's dominant culture and language is a priority. An expected result of thoughtful antiracism training would be to challenge this lack of basic equity—a language barrier—as a simple first step to achieving understanding.

Growing Power has been able to create measurable effective methods and appropriately scaled urban and community sustainable food systems outside an academic institution. We developed food systems reflecting an understanding of agronomy, soil science, aquaponics, and regenerative education within a series of production environments. These systems and methods are finally being validated by institutions of science, even though other measures have already shown that they clearly work. This process can be exhausting and frustrating for all parties, but defining why this cultural gap exists is the beginning of undoing structural, institutional racism and creating true equity.

FUNDER AND COALITION INSIGHTS INTO SOCIAL EQUITY AND JUSTICE

I (Rodger) have been engaged for sixteen years in community food systems, urban agriculture, community development, and, increasingly, equity and antiracism efforts. I started working for an international funder in support of urban agriculture and small-farm projects in the Midwest, then launched the International Network for Urban Agriculture and taught urban food systems at the Illinois Institute of Technology. I cofounded the Chicago Food Policy Action Council in 2001, cochaired it for a long time, and am now a board member and coordinator.

As a white male I have found the aforementioned experiences to be incredibly meaningful, frustrating, and humbling. Most of my efforts have been focused in Chicago, an amazing city where almost every American issue of race, justice, and equity is in play. Nothing is easy there, and I increasingly recognize that the challenge has to do with underlying power issues.

I continually struggle to understand the differences between the words, actions, intentions, and histories of players and stakeholders in the community food systems arena. As a funder I found it challenging to grasp who the real community representatives were—not the front people who learned the language that funders want to hear. In grant solicitations and proposal requests, funders use terms and language to indicate what they are looking for, but these terms can be loosely defined. Here are some examples:

- *Community, community-led, community-driven.* Who makes up a community, who has leadership, and who makes decisions?
- *Sustainability, triple bottom line.* What are the measures, and who sets the standards, for determining economic success (return on investment?), environmental success (more organic produce?), and social sustainability (high school graduation rates?)?
- *Community food systems, food security, food access.* What is prioritized: access to calories, culturally significant foods, fresh produce? Do these phrases include the related issues of education, violence, public safety, and affordable housing?
- *Scalability, replicable models.* How do we determine, and who decides, what is the "right" model to apply in different cities and neighborhoods?

In my different roles I have seen how organizations learn to use the language and culture of funders to gain access to resources while not challenging the system that reinforces status quo inequities. Organizations that are successful at acquiring the funders' resources may or may not have effective programs, but they are good at adopting the language and jargon. Resource providers, including funders, public agencies, and universities, can set the terms for much of the supported work in communities, but this effort often does not include changing the underlying issues.

As a funder I knew the contortions that community partners must go through to fit into our frameworks. The following examples highlight some of the practices that challenge community food system change, particularly for organizations with limited resources led by people of color:

- Most grant resources tend to be directed through white-led intermediaries before getting to the community.
- Conversations among funders and other power networks may exclude potential grantees who are publicly critical of the status quo and who risk

being labeled as difficult to work with and unworthy of funding.

- Potential grantees must deal with multiple layers of gatekeepers, including consultants, rotating staff members, and funder consortiums.
- Funders have unique applications and processes, limited collaborative funding, different funding cycles, and shifting priorities. The public information on these priorities can be vague and confusing.
- Grant exclusions by funders, including strict support limits on expenses such as capital acquisitions, advocacy or policy activities, personnel, and overhead, require organizations to spend precious time and energy piecing together funding sources to cover their expenses.
- Funders may emphasize the importing of external solutions rather than community-developed goals. For example, a primary fix proposed for food deserts in many cities is to offer incentives to the same large supermarkets that originally abandoned the areas.
- Information about opportunities and access to decision makers is often shared during in-person meetings, which requires staff members or representatives from community organizations to be in the room. This can be challenging for organizations with tight resources that cannot spare the hours required for travel and meetings.

Since 2000 many more funders, institutions, and organizations have become involved in the sphere of community food systems, but there remains the fundamental issue of how to move away from a focus on emergency services and piecemeal project funding to one that addresses underlying systemic issues with a clear understanding of the inequities caused by institutionalized racism. I have participated in several antiracism training sessions over the past few years and have gained a new understanding each time of the institutionally racist food system, while also examining my own white privilege and standing up as a white ally.

The following language can provide, through a shared understanding of the definitions, a clearer vision for how change can occur in community food systems:

- *Equity*: equal access to resources in and for communities of color and perhaps additional investment to address inequity.
- *Justice*: intentionally addressing inequity in institutions, systems, and policy.
- *Antiracism*: recognizing the existence of institutional racism, white privilege, and internalized oppression.

Institutions that support systemic change should challenge themselves deeply and develop shared definitions of key terms with the communities they wish to support. Funders and other institutions must understand how they set barriers that impede results in communities of high need. They should take chances on emerging organizations and leaders in historically undersupported communities, recognizing that a wide range of success is possible and that it takes consistent, long-term investment to deal with decades of noninvestment in the physical, economic, educational, and health systems in communities of color. It is also vitally important for funders to recognize the role of policy and advocacy in creating more lasting systemic change.

ACKNOWLEDGING PRIVILEGE AND POWER DYNAMICS IN CREATING EQUITABLE COMMUNITY PARTNERSHIPS

Understanding white privilege uncovers the understanding of power dynamics. In working as a white woman within communities of color throughout Chicago, I (Laurell) have learned the effects of external oppression as well as the internal dynamics of staff members and partners of color who have experienced internalized oppression. It is important to acknowledge that privilege happens on many levels: societal, organizational, and individual. More equitable community partnerships may develop as white people and white culture seek to acknowledge their privileges and power.

Language continues to be important, as does cultural dissonance, as we shift our attention to discussions of privilege. When the south and west sides of Chicago are covered in the media, for instance, they are often described using infamous terms such as *Chiraq* and *food deserts*. Chiraq, a portmanteau of *Chicago* and *Iraq*, is used as code to refer to the violence and poverty of these neighborhoods. What is often underreported is the complexity of these neighborhoods, which constitute two-thirds of Chicago's population. Food justice work in Chicago has revealed real tales of families working hard to provide better lives for their young and of urban farmers trying to provide healthy food to areas that grocery stores refuse to serve because of perceived violence and poverty. Nonprofit organizations dedicated to food justice try to improve education standards and services for people of color, who have less access to resources than the predominantly white citizens on Chicago's north side.

Many nonprofit organizations or institutions that work in communities of color are run by well-meaning white people who want to help but who are often perceived as experts or saviors. Some enter neighborhoods with resources, an intention to serve the constituents, and an agenda for getting the job done. More often than not, however, their action plans ignore or limit community input. In my opinion there are three overarching reasons why "serving" undermines the very outcome we often strive to achieve as white people working in communities of color. First, we do not account for how racism affects people of color or for the power imbalances and privilege dynamics between institutions and community partners. Second, imposing a top-down solution onto black and brown communities often undermines empowerment. Third, "partnerships" between institutions and community groups are unequal if the "service" is created not out of humility but from a sense of superiority.

Experience shapes our perception of how we walk through the world. On an individual level we each carry a degree of privilege, which can manifest itself in a variety of ways. If we imagine privilege in terms of a racetrack with a starting gate and a finish line, for some the track is empty and the run is smooth and swift. For others the track looks more like an obstacle course, with seemingly insurmountable hurdles blocking the way. These hurdles may include poverty, lack of access to nutritious food, underfunded schools, police harassment and brutality, domestic violence, sexual assault, mental health issues, and substance abuse. No person is automatically immune to these issues, but when you add overt racism and microaggressive racist behaviors—to which no person of color is immune—it makes the weight of overcoming those hurdles much heavier. Racist acts, whether unintended slights or overt aggression, affect a person's mental health. They are an enormous burden to carry, and when combined with poverty, violence, and lack of resources, they make an entire community particularly vulnerable.

When white-led organizations come into communities of color with agendas and game plans, those plans often do not address the root causes of racism. The agendas tend not to include community input, and instead they position the outside institutions in the roles of expert and savior, which reinforces institutional racism. The game plans are almost always rooted in privilege. For example, an early-morning meeting in a downtown high-rise to discuss progress does not take into account that the community partners might not have access to transportation or child care or may be uncomfortable being greeted by a security guard in the lobby. In addition, resources may be withheld if timelines aren't met or if participants cannot meet certain standards, but organizations don't take into account the structural or institutional hurdles that have not been addressed or improved in the first place. Once the funding for the project runs out, the institutional partner often abandons the community partner, creating a void of resources, damaging community trust, and making it harder to create just and equitable partnerships in the future. Privilege creates blind spots in transactions with an individual community, and unless that community is treated as an equal partner in devising solutions and addressing the root problems of its own neighborhoods, the cycle of power and oppression will continue.

Another downside of a lack of community input and involvement occurs when the "expert" institution uses a generic method intended to work universally rather than tailoring a method to address the particular needs of a community. For example, a nonprofit organization or institution may come into a neighborhood and establish a weekly farmers' market that provides residents with fresh produce. However, what the organization or institution did not take into account is that the community had been advocating for a neighborhood grocery store. The farmers' market, while beneficial, is not daily or year-round, and considering that only 2 percent of farm owners in this country are people of color, a farmers' market is also not representative of the community. So instead of helping to establish what the community wants and empowering it to have access to fresh produce throughout the year, the institution created a Band-Aid solution to which many in the neighborhood feel a lack of connection or are afraid or unable to access. Even though the effort was certainly well-meaning and better than the status quo, working with community partners and listening to what they want would have helped empower neighbors, as opposed to implementing a top-down solution that did not respond to the community's needs.

Top-down solutions with limited or no community input reinforce the idea that the community partners' needs and desires are not important. This model establishes a power shift that puts the outside institution into the role of expert and the community partner in the role of beneficiary, despite the fact that the community partner is often doing most of the work. In this role the institution often speaks on behalf of the community partner, does not value the partner's schedule or workload, and can threaten to withhold resources as a form of currency—if the community partner can't meet the institution's deadlines, it is replaced with another entity. Such manipulation continues the cycle of institutional racism and gatekeeping, holding communities hostage to top-down institutional demands. This behavior also pits community organizations that should be working together against one another in a competition for resources.

In order to be effective allies, white-led organizations need to treat community groups as true partners with a voice and a vote in planning and decision making. Moreover, meeting the community residents where they live, employing them, training staff members on dismantling racism, and being transparent at every step of the process are necessary objectives.

The authors thank Malik Yakini for his contributions to this chapter.

CHAPTER FOURTEEN

Innovations and Successes

STEVE VENTURA

In this chapter the codirector of the CRFS project summarizes each of the seven CRFS values and provides additional observations and project anecdotes.

I struggled for many weeks with how to orient and write this last chapter. I knew the purpose: to summarize key points and briefly describe significant activities and findings not covered elsewhere. I even had a structure: the seven values of the systems framework wheel shown in figure 2. But I couldn't figure out how to convey these ideas with the dispassionate third-person voice that I use for academic writing. Although this voice can be very useful for conveying facts and describing activities, it does not convey the full scope of community and regional food systems and the people working to make them better. Their efforts and interactions, and their struggles and hopes, are better conveyed as stories and examples. My "aha" moment came when I recognized that my writing should include some of this drama. Perhaps to the chagrin of editors everywhere, this chapter blends the objective and the emotional, in more of a blog voice than a professor's voice. Here are some of the facts and the stories, organized around the values identified by the CRFS project (and shown as subsections in this chapter), that drive community and regional food systems.

VIBRANT FARMS AND GARDENS

Will Allen says that his headquarters operation in Milwaukee—old floral greenhouses retrofitted for urban agriculture production—generates five dollars per square foot of profits. A good yield for a Wisconsin corn farmer is two hundred bushels per acre; in March 2016 Wisconsin corn farmers were getting about $3.25 per bushel, or about 15 cents per square foot. The difference is a result of both the commodities involved (e.g., sprouts and fresh greens versus corn) and the space-intensive nature of urban gardening and farming. In an extension of his current vertical farming (closed-system aquaponics with layers of green plants atop fish tanks), Will promotes his vision of a vertical farm as layers of a building that could grow five times what his greenhouses produce in the same floor space.

This orientation to space-intensive and season-extending production epitomizes urban agriculture. It is manifested in other forms as well, such as rooftop gardening and hoop houses. Stephanie Calloway, a coauthor of chapter 10, tends a "healing garden" for CORE/El Centro on the roof of its Milwaukee building. Growing Power is working with Milwaukee Public Schools to build hoop houses on three campuses, which will provide food growing spaces and jobs for young black men in an area with an alarmingly high unemployment rate.

The need to grow a lot of food in relatively small spaces is a unique aspect of urban agriculture. It is a response to challenges related to land access and land tenure security, as discussed in chapter 2, and the need for sustainability. One of the aspects of sustainability is profitability of the operation; *Whole Measures for Community Food Systems* (Abi-Nader et al. 2009), the document that guided much of the CRFS project's values-based approach to food project innovation and success, declares that a vibrant farm "supports local, sustainable family farms to thrive and be economically viable." A bad old joke about farming starts by questioning a lucky farmer, "What are you going to do now that you've won a million dollars in the lottery?" He replies, "Well, I reckon I'll keep on farming till the money runs out." Although the *Whole Measures* definition of vibrant farms suggests that a community food system has to generate income to be viable, it does not mean that a farm or garden must provide the entire livelihood for a farmer. Even rural farms often have other, off-farm income, and in a city farming is typically a supplement to income rather than a sole source. Therefore, it's important to look at all the values associated with community food systems; for some growers, generating just enough cash to break even may be offset by many other positive aspects.

Hallway conversations at food system conferences include discussions of the imperfect triad of urban farming: you can grow a lot of food in a small space and make money, or you can grow a lot of food in a small space and generate community services, but it's very hard to do all three. A survey by Dimitri, Oberholtzer, and Pressman (2016, 603) found that "urban farms, regardless of mission, are relatively small and face similar challenges in terms of providing the primary farmer with a living." Of course, this oversimplifies a complex equation, but the gist of it seems to hold true in many situations. As noted in chapter 3, this dilemma is significant for community organizations with primary missions in community services that add urban production to their portfolio without a skilled farmer on staff. Even Growing Power, which does generate revenue from its production, depends substantially on grants and contracts for its community outreach and service.

Another way to build the triad is through public support of urban farms and gardens. For more than eighty years, federal policy has provided commodity payments, crop insurance, and other forms of support for traditional rural agriculture. These programs are justified for national food security—a safe and affordable food supply for the nation—and for the continued viability of the agricultural sector and livelihoods of farmers. Urban agriculture also generates valuable social services, which is starting to be recognized in federal policy. In the last decade, which encompasses the 2008 and 2014 Farm Bills, several new policies and programs have developed provisions that can be used by urban producers. Growers may be able to take advantage of loan, market development, and other grant programs that target local and regional food systems. Some programs specifically target beginning, socially disadvantaged, and veteran farmers. A few of the conservation programs can support urban agriculture; for example, the Environmental Quality Incentives Program has been used to build hoop houses in Milwaukee. The National Sustainable Agriculture Coalition maintains up-to-date Internet resources about federal programs and periodically publishes guidebooks (Fitzgerald et al. 2010). The coalition worked with the Michael Fields Agricultural Institute on a guide to federal programs (Krome and Reistad 2014), available through the National Center for Appropriate Technology. The USDA (2016) provides its own resource guide on urban agriculture.

State and local governments may also provide resources that urban growers can tap into. Many states have farm-to-school programs and local-food initiatives. As discussed in chapter 2, one of the frontiers in public subsidies is working with state and local property tax authorities. Many jurisdictions tax rural farmland at rates proportional to use value for agriculture rather than the full market value. If this concept can be extended to city land, where market values for land are very high, it would substantially alleviate the cost of farming.

Any conversation about food security in the United States must include race as a central component. This book has two chapters primarily focused on the experiences of African Americans, and evidence of systemic racism's effects on the food system is found in the other chapters. Other minority communities, particularly those characterized by low incomes and high unemployment, also suffer from indifferent and exploitative food systems. Power structures reinforce long-standing social and cultural prejudices that are rooted in racism. For example, black farmers had to sue the federal government to recover damages for decades of bias in federal loan and commodity programs. Latino immigrants who fear reprisal for their undocumented status are ruthlessly exploited by the food processing industry. Structural and institutional racism is built into many parts of the food system, particularly in urban areas, where lack of access to good food, clean water, and safe housing are often tied to race and class.

When we, mostly white academics, started the CRFS project, one of our community partners insisted that anyone it worked with should first undergo training in dismantling racism. This was initially perceived as an unusual request, because most of our group had already acknowledged issues of race in the food system and was familiar with the need to honor and incorporate experience and local knowledge in our activities. However, we did participate in antiracism training through the Growing Food and Justice for All Initiative at the Growing Power Urban and Small Farms conferences and in the initiative's workshops (Intensive Leadership and Facilitator Training: Dismantling Racism through the Food System) at the Growing Power Iron Street Farm in Chicago.

These training sessions clearly had an impact on how the CRFS project evolved. We made more efforts to hire people of color, and we openly discussed the important process of recognizing differences in privilege and reducing its influence. I believe and hope that we became better at listening, empathizing, sharing power, avoiding microaggressions, and learning many other lessons. Whites working to dismantle structural racism find that it is complicated and at times uncomfortable, but the need to be open and purposeful about it is apparent. Part of my standard pitch to academic colleagues now interested in food system work is to start a project exactly where we were asked to start with the CRFS project: with better understanding *and feeling* for the effects of racism in food systems.

Our original project proposal included a plan to go into communities and conduct research. Even in that approach it is necessary to respect local culture, history, and, most important, people. People of color in the United States have experienced long histories of exclusion and exploitation, which includes researchers coming into communities and leaving—"helicopter research" that is typified as contributing nothing of value to the communities themselves. We did not want to replicate the same oppressive role that white academics have implemented in such communities for decades, so we quickly abandoned case study research. Instead, our ten Innovation Fund projects, most of which are described in this book, were entirely devised and led by community organizations, and our scores of community engagement projects started by listening to what community organizations and leaders wanted and determining how we could support their goals.

As described in chapter 9, white allies in Detroit's food system have a particular role to play in creating equity. In chapter 13 our partners in Chicago describe similar challenges arising from structural and institutional racism in food systems. Most large cities have organizations with antiracism programs that can be a good starting point. We found the Growing Food and Justice for All Initiative particularly useful because of its orientation to food systems; I recommend its readings and have included its website address in the references.

Of course, one of the most important ways to make food systems fairer and enable local economies to thrive is to make sure that everyone involved is paid a living wage. Although that subject is beyond the scope of this book, it is encouraging to see local initiatives to raise minimum wages and scary to see states enact preemption laws that undermine local efforts (one more reason to build strong communities!).

STRONG COMMUNITIES

With perhaps the exception of hunter-gatherer tribes, communities do not function primarily around food. However, food is deeply imbued in cultures and is something we all relate to in a personal way. Organizing around food and food systems can be a powerful means to uplift communities: revitalizing downtrodden neighborhoods and repurposing derelict land, providing employment opportunities, educating and engaging youth, reducing crime, and providing local fare on tables. Without exception, meetings and events are better with food, particularly if it's local food with a story about the participating farms, growers, processors, or chefs. Although bureaucracies can make provisioning with local food difficult—for example, requiring food vendors to be certified by a public agency—it is worth the effort, both for the vendor and for the community building it engenders.

Walnut Way Conservation, an established Milwaukee neighborhood organization, epitomizes this way of using food as part of a strategy for strengthening its Lindsey Heights community, an area in north-central Milwaukee once known for drugs, crime, and violence. Community and backyard gardens and improved access to food are part of a comprehensive approach that includes promoting civic and community leadership, housing rehabilitation, classes and training, and local economic development. The culmination of the local economic development was the opening of the Innovations and Wellness Commons in 2015, which attracted a local grocery cooperative, a juice bar, and office space for local businesses. Walnut Way's efforts directly respond to community food needs and effectively encourage civic participation and local political leadership. The activities are summarized in this snippet of Walnut Way's history from its website (no longer available): "Walnut Way residents and volunteers have five years of successful experience in urban ecology-based initiatives, including nearly eliminating drug and prostitution activity in the neighborhood; creating and managing multiple, high-production community gardens; conducting successful, profitable sales of garden produce; ongoing gardening and nutrition education programs for youth and adults . . . and converting a former drug house/murder site into a prime turn-of-the-19th century restoration which will serve as a neighborhood gathering spot for educational as well as social purposes."

The CRFS project made only a small contribution through an Innovation Fund award to Walnut Way's overall effort, helping with a vegetable washing station and corner store initiatives. But our project mission has been to identify innovations and promote successes, and this is a clear example of both.

THRIVING LOCAL ECONOMIES

In my first draft of this chapter, I started this section with the frequently used derogatory term for economics: "the dismal science." When I looked up the origins of the term online, I was amazed that its derivation is surprisingly relevant to food system work, at least according to Derek Thompson (2013), a writer for the *Atlantic*. It seems that Thomas Carlyle, the man who coined the term early in the nineteenth century, was writing about slavery in the West Indies. He called economics dismal because it couldn't offer support for his proslavery beliefs. Thompson noted that the term "aligns economics with morality, and against racism." Local food system work has the potential to encourage thriving local economies. In order to demonstrate this potential, we delve into the dismal science—or maybe it's the black arts—of economics.

A tool that is commonly used to convey the value of community food systems is cost-benefit analysis (CBA). As part of the CRFS project, Jenny Buckley and Chris Peterson (n.d.) of Michigan State University developed an introduction to CBA specifically within the context of urban agriculture. As noted in the preface, CBA "can provide a powerful tool for communicating the economic value of urban agriculture to policymakers, funders, and other decision makers. This guide provides an introduction to cost-benefit analysis and helps urban agriculture practitioners develop a plan for conducting a preliminary-level CBA."

CBA and related economic approaches may be used for evaluating individual projects or proposals and for measuring or predicting the community-level effects of food system policies and programs. The guide focuses on benefits to community economic development, human development, neighborhood revitalization, food security, and environmental improvement, and it discusses calculating costs.

To evaluate the benefits, the guide suggests identifying *surrogate measures* of benefits. The idea of surrogate measures is that many of the values of community food systems are not ordinarily measured in monetary terms, so we find approximations that serve for CBA purposes. For example, we can't directly measure children's increased ability to focus on their schoolwork when they aren't hungry, but we can measure the number of children who participate in free and reduced-price meal programs. Local job creation and multipurpose infrastructure development are other typical examples of functions that we try to monetize for CBA.

Readers who want to further apply CBA to community food systems will probably use the following concepts:

- **Leakage.** This refers to food dollars leaving the local economy. Kelly Cain, a colleague at the St. Croix Institute for Sustainable Community Development, estimated that 90 percent of the money spent at local stores and restaurants left their northwest Wisconsin counties, even though these counties are in the midst of an agricultural area. Retaining even a small portion of this loss could have significant value to the local economy and area residents.
- **Multiplier.** This refers to the extra value of money kept in the local economy. A local grower who gains extra cash through value-added processing of his or her crop can spend it in the local hardware store rather than having it end up in distant corporate headquarters.

- **Opportunity cost.** This is the value of alternatives to a given activity. Alternatives might also be analyzed through CBA. So, for example, we could compare the value of converting vacant lots to community gardens versus converting them to baseball diamonds, at least in monetary terms.
- **Externalities.** These are costs and benefits that are not incorporated in an analysis, although they are recognized as part of the overall system. On the negative side, industrialized agriculture would consider water quality degradation or worker pesticide poisoning as an externality of the production system. A positive externality of urban agriculture might be reduced urban storm water runoff. In both cases these factors may not be included in a CBA, but they should at least be explicitly recognized as parts of the system.

Not all community food system activities have to contribute directly to local economies, but successful local businesses can create a momentum that, in aggregate, does provide for economic resilience and vitality. One potential externality, however, is the process of gentrification, which has both positive and negative aspects. We hope that tools such as CBA will allow food system activists, entrepreneurs, and policy makers to keep neighborhoods affordable even as they are upgraded.

The Multiplier Effect

STEVE VENTURA (ADAPTED FROM
BAKER AND THOMAS N.D.)

It is a cold winter day in a little Wisconsin village. The snow is blowing, and the streets are deserted. Times are tough, everybody is in debt, and everybody lives on credit. A rich tourist from the big city stops at the local bed-and-breakfast and lays a $100 bill on the desk, telling the owner he wants to inspect the rooms upstairs and then pick one in which to spend the night.

The owner gives him some keys and, as soon as the visitor has walked upstairs, grabs the $100 bill and runs over to the market to pay his debt. "Business is slow," he tells the market owner. "Thanks for the food you fronted me."

The market owner takes the $100 bill and runs down the street to repay his debt at the hardware store. "Thanks for fixing my freezers the other day," he says.

The hardware store owner has run up quite a tab at the tavern, so he takes the $100 bill and heads off to pay his debt there, exclaiming, "I hope I'm still welcome here now!"

The bartender then rushes to the bed-and-breakfast and pays off her room bill to the owner with the $100 bill. "Thanks for putting up my sister-in-law last week," she states.

The bed-and-breakfast owner places the $100 bill back on the counter so the rich traveler will not suspect anything. At that moment the traveler comes down the stairs, states that the rooms are not satisfactory, picks up the $100 bill and pockets it, and leaves town.

No one produced anything. No one earned anything. However, the whole village is now out of debt and looking to the future with more optimism.

SUSTAINABLE ECOSYSTEMS

The usual way of describing sustainable agriculture is as a system that is environmentally sound, socially acceptable, and profitable. When done well, however, community-based agriculture should be able to go well beyond just sound and acceptable. (The issue of profitability was discussed in the "Vibrant Farms and Gardens" section earlier in this chapter.) The Millennium Ecosystem Assessment (MEA; 2005), an international effort to assess the health of the planet, popularized the term *ecosystem services* to connote the environmental outcomes of human activities and defined it in four broad categories of services: provisioning, regulating, cultural, and supporting. In examining community and regional food systems contributions, we have found that certain food system social services provide parallels in all four categories.

Here are some examples of the ecosystem services of community and regional food systems:

- Provisioning: Provides good food with minimal environmental degradation.
- Regulating: Contributes to storm water management (reduced runoff) in urban areas.

- Cultural: Creates opportunities for outdoor recreation and education.
- Supporting: Recycles food waste into soil and plant nutrients.

Here are some examples of the social services of community and regional food systems:

- Provisioning: Provides opportunities for employment in high unemployment areas.
- Regulating: Contributes to neighborhood revitalization and crime reduction.
- Cultural: Ties people to cultural roots through heritage gardens and food ways.
- Supporting: Keeps capital in communities.

MEA (2005) describes "constituents of well-being" that are quite similar to the terms we heard regularly from our project partners: "freedom of choice and action, security, basic material for good life, health, and good social relations." These broad concepts, intended to characterize biodiversity and life on earth, are also clearly applicable to the specific context of food system sustainability. A challenge ahead for food system academics and activists will be to develop and validate measures to represent these constituents. Some measures, such as the amount of food produced from a farm or a garden or the amount of food waste put into a compost stream, will be relatively straightforward. Other aspects, such as neighborhood character or changes in food awareness, will be much more challenging to measure. In both cases this will be important evidence to justify policies and investments that support community and regional food systems.

The growing peril of climate change means that people who are engaged in creating sustainable food systems will also have to be cognizant of the role of food in greenhouse gas production and the need for mitigation and adaptation as the climate changes. Depending on the assumptions about what is included in each of the following categories, various estimates attribute one-quarter to one-third of all human-created greenhouse gases to food production, processing, distribution, preparation, and waste management. Shifting a portion of food system activities to the kinds of approaches described in this book would have an almost imperceptible effect on greenhouse gases on a global scale. But changing how we think about food can be part of a growing awareness of the need to change many things that affect greenhouse gas production, both in our personal lives and in broader corporate, institutional, and political decisions.

Learning to live with climate change entails two broad strategies: adaptation and mitigation. Climate change will be marked by increases in the average global temperature *and* by more extremes in weather: longer droughts and floods, more extreme temperatures, and bigger events like tornadoes and hailstorms. Crop diversity, as commonly practiced by CRFS project growers, is a clear adaptation strategy. When one crop fails because of extremes, another may flourish. Hoop houses and other structures make longer growing seasons even longer. CRFS project growers are often experimenters, figuring out what works in changing conditions.

Mitigation entails efforts to reduce the rate of climate change. Because fossil fuel combustion for transportation contributes to greenhouse gases, the local food movement has focused on the concept of food miles (Paxton 1994) to measure the greenhouse gas contribution of foods—their carbon footprint, as it is sometimes called. A growing body of scholarly work shows that transportation is a relatively small portion of the overall effects of agriculture, but it can be a useful point of comparison and help people consider the source of their food. For example, a customer in Chicago might assume that an apple grown in Wisconsin has less of a carbon footprint than one from Argentina. But even this isn't as simple as it sounds. If the Wisconsin apple was purchased in the spring, it probably came from a storage facility that depends on fossil fuels and was transported in a small fuel-inefficient truck. If the apple came from Argentina, it probably came fresh from the field on a very fuel-efficient boat.

This complexity is part of the reason we chose to refer to community and *regional* food systems. A region can be defined for each type of food that's grown. For highly perishable products like microgreens or heavy products like strawberries, an efficient and appropriate source region might be quite small. For example, microgreens or mushrooms can be produced right in Milwaukee for Milwaukee consumers. For other products that are stable, easy to ship, and most efficiently produced in large quantities through highly mechanized farming, the appropriate region may be much larger. In Wisconsin, potatoes and green beans can be grown very efficiently in the Central Sands region, more than one hundred miles from Milwaukee. This process generates less greenhouse gas and provides cheaper food for those in need. This is not to imply that some consumers won't demand organic potatoes and that growers in and near cities shouldn't grow them; diversity in our food system is good! The definitions and criteria for appropriate and sustainable regions, or foodsheds, will continue to be researchable and debatable questions.

As noted in chapter 1 when explaining the evolution of our food system framework, we purposely did not include how food choices and food behaviors affect personal and public health. These are important issues and are definitely part of the overall food system framework. But they are being pursued by many skilled researchers through projects funded by millions, if not billions, of dollars. We believed it would dilute the focus of our project and require different participants and collaborators to pursue these questions in depth.

Nonetheless, we know from our community interactions that the potential for improving health motivates many to become involved in food system change. And we know from countless studies that the maxim "you are what you eat" is true: the kinds of food we eat do affect our health. From this perspective, our project was built around a big assumption: if healthy food is available and affordable, people will make healthier choices. Unfortunately, as noted in chapters 5 and 7, this doesn't always follow. Habits are hard to change, junk food is tasty, and not everyone has stoves, utensils, and the skills to prepare food. Changing eating behaviors is a challenge that will play out in many ways, with different strategies required in different communities.

One approach to changing eating behaviors is to deliver the message directly to people where they acquire food. On Milwaukee's south side, where many of the city's Latinos have settled, we promoted the beginning of what today is the Southside EAT (Eating, Access, and Transformation) Coalition. As project collaborator and community organizer Tatiana Maida reports, EAT has the goal of "inspiring healthy food choices and behaviors by enhancing the food environment and celebrating the diversity of Milwaukee's south side." Four organizations were part of this coalition: the Sixteenth Street Community Health Centers, CORE/El Centro, the Center for Urban Population and Health, and UWEX. The first project they worked on together was the Healthy Grocery Store Campaign, an idea of the community group Latinos por la Salud to increase the number of healthy food options in the neighborhood's two main grocery stores. Latinos por la Salud and EAT worked together with Pete's Fruit Market and El Rey Supermarket to introduce hormone-free milk, antibiotic-free eggs, quinoa, sesame seeds, flaxseed, and bread and cereals made without high-fructose corn syrup.

An interesting aspect of this campaign was the difference in position between dietitians from UWEX and the community group. Whereas the dietitians opposed focusing on products without high-fructose corn syrup ("sugar is sugar"), the community members said that beyond the controversy about corn syrup in the nutrition field, families needed to have better options in their stores. Given that this campaign was initiated by the community group and the goal of EAT was to support and enhance their efforts, not change them, the community request prevailed and was communicated to the stores. A marketing campaign—along with "shelf talkers" (in-store signage), cooking demonstrations, and in-store education by community health promoters—was used to help customers learn about the new items. Today these items are sold in the stores, and south side families no longer have to go to other neighborhoods or natural food stores in search of these products.

The question of eating behavior is a subset of a broader issue: the challenge of attributing specific health benefits to food system activities. In the jargon of medical research, do food system interventions result in measurable health outcomes? This challenge requires identifying the right variables and measurements of food system changes—for example, changes in the availability of fresh fruits and vegetables coupled with the identification and measurement of potential changes in health at a personal or population level. Establishing convincing evidence of a causal relationship between the two is a fundamental challenge. Many other factors influence health, and people don't readily recall or record everything they eat; many competing variables thus confound the relationship. It will be useful to know which food system activities and changes have positive effects on health, but this is a long-term proposition that requires many kinds of evidence. As explained in the next section, we hope and believe that the new Cooperative Institute for Urban Agriculture and Nutrition in Milwaukee (see chapter 10 and below) can contribute to that evidence.

SYSTEM THINKING AND COLLABORATION

The last value of our community and regional food systems framework shown in figure 2 is system thinking and collaboration. This was *not* part of the *Whole Measures* (Abi-Nader et al. 2009) fields and practices, although it is certainly implied by many of the suggestions. Indeed, few efforts are ever likely to succeed without thinking broadly, collaborating, and communicating. Chapter 13 describes the unique challenges of communication across worldviews—between academics and practitioners, particularly in small, nonprofit community organizations. In summary, this communication requires humility, empathy, and a willingness to listen and learn on the part of all parties. Chapter 10 covers the principles and practices of collective impact. Although the term is becoming overused, it still quickly and cogently describes what participants in food system change believe is necessary to make progress.

I believe and hope that one of the enduring legacies of the CRFS project will be support for the creation of the Cooperative Institute for Urban Agriculture and Nutrition (CIUAN) in Milwaukee. In recognition of rampant food insecurity in Milwaukee and related problems of hunger and public health, a broad cross-section of public and private interests formed a coalition in 2012. A memorandum of understanding was signed by seven academic institutions, the City of Milwaukee, Growing Power, and the Milwaukee Food Council, which represent many community organizations in urban agriculture, food systems, and public health. This agreement provided a starting point for raising funds, engaging additional partners, and creating a formal structure to maintain the coalition over time. In the summer of 2016, papers were filed with the State of Wisconsin to formally charter CIUAN under Wisconsin cooperative law. This unique institutional structure provides a means for universities and colleges, public agencies, businesses, and nonprofit organizations to work together on the complex challenges of food systems. CIUAN will harness and align the research capacity of universities with the needs and front-line skills of agencies and organizations, particularly through needs identified by close engagement with the Milwaukee Food Council. CIUAN will also address food security through collaborative curriculum development and delivery, intern placement, technical assistance and training for growers and entrepreneurs, policy advocacy, and the facilitation of connections among many sectors of the Milwaukee area food system. Although it remains to be seen how much of this ambitious agenda is realized long-term, the prospects are bright.

The purpose of this book was summarized by the tagline of our project: to identify innovations and promote successes. Both of these activities will continue to be challenges for food system researchers and practitioners long after this book has gathered thick dust on a bookshelf. It is difficult to know whether we've helped to move the situation very far in a positive direction, toward what Will Allen has termed "the Good Food Revolution," but I hope we have provided some ideas and inspiration to advance our cause. If you've made it this far through the text, you must care, so keep up the good work!

References

BOOKS AND ARTICLES

Abi-Nader, Jeanette, Adrian Ayson, Keecha Harris, Hank Herrera, Darcel Eddins, Deb Habib, Jim Hanna, Chris Paterson, Karl Sutton, and Lydia Villanueva. 2009. *Whole Measures for Community Food Systems: Values-Based Planning and Evaluation.* Vayston, VT: Center for Whole Communities. http://foodsecurity.org/pub/WholeMeasuresCFS-web.pdf.

Aldrich, Rob, and Melissa Levy. 2015. "Assessing, Planning, and Measuring Community Conservation Impact: A Tool for Land Trusts." Land Trust Alliance, http://www.landtrustalliance.org/publication/community -conservation-tool.

Alkon, Alison Hope, and Teresa Marie Mares. 2012. "Food Sovereignty in U.S. Food Movements: Radical Visions and Neoliberal Constraints." *Agriculture and Human Values* 29 (3): 347–59.

Allen, Will. 2012. *The Good Food Revolution: Growing Healthy Food, People, and Communities.* New York: Gotham Books.

American Planning Association. 2007. "Policy Guide on Community and Regional Food Planning." https://www.planning.org/policy/guides/ adopted/food.htm.

———. n.d. *Planning a Healthy, Sustainable Food System.* https://www.planning .org/nationalcenters/health/pdf/apapchfoodsystemplanning.pdf.

Anderson, Jason. 2005. "Counties, Cities, Villages, Towns: Forms of Local Government and Their Functions." Wisconsin Legislative Reference Bureau, http://legis.wisconsin.gov/LRB/gw/gw_6.pdf.

Baics, Gergely. 2009. "Feeding Gotham: A Social History of Urban

Provisioning, 1780–1860." PhD dissertation, Northwestern University, Evanston.

Baker, Will, and Thomas, Guy. n.d. "Funny Economic Jokes." Funny Jokes, Amusing Pictures and Videos, http://www.guy-sports.com/funny/funny _economic_jokes.htm.

Barling, David, Tim Lang, and Martin Caraher. 2002. "Joined-Up Food Policy? The Trials of Governance, Public Policy, and the Food System." *Social Policy and Administration* 36 (6): 556–74.

Barton, K. L., W. L. Wrieden, and A. S. Anderson. 2011. "Validity and Reliability of a Short Questionnaire for Assessing the Impact of Cooking Skills Interventions." *Journal of Human Nutrition and Dietetics* 24: 588–95.

Bell, Judith, Gabriella Mora, Erin Hagan, Victor Rubin, and Allison Karpyn. 2013. "Access to Healthy Food and Why It Matters: A Review of the Research." PolicyLink and the Food Trust, http://thefoodtrust.org/uploads/ media_items/access-to-healthy-food.original.pdf.

Bentley, Stephen, and Ravenna Barker. 2005. *Fighting Global Warming at the Farmers' Market: The Role of Farmers' Markets in Reducing Greenhouse Gas Emissions.* Toronto, ON: FoodShare Research in Action.

Bittman, Mark. 2015. "A Rare Pleasure: It's Easier and Cheaper Than You Might Think to Play Steakhouse Chef for a Night." *New York Times*, August 18.

Bittner, Jason, Lindsey Day-Farnsworth, Michelle Miller, Rosa Kozub, and Bob Gollnik. 2011. *Maximizing Freight Movements in Local Food Markets.* Madison: University of Wisconsin.

Block, Daniel, and Joanne Kouba. 2006. "A Comparison of the Availability and Affordability of a Market Basket in Two Communities in the Chicago Area." *Public Health Nutrition* 9 (7): 837–45.

Bloom, J., and C. Hinrichs. 2010. "Moving Local Food through Conventional Food System Infrastructure: Value Chain Framework Comparisons and Insights." *Renewable Agriculture and Food Systems* 26: 13–23.

Born, Branden, and Mark Purcell. 2006. "Avoiding the Local Trap: Scale and Food Systems in Planning Research." *Journal of Planning Education and Research* 26 (2): 195–207.

Borron, Sarah Marie. 2003. *Food Policy Councils: Practice and Possibility.* Eugene, OR: Congressional Hunger Center.

Boston Redevelopment Authority. 2013. "Draft Article 89 as of 5.22.13." http:// www.bostonredevelopmentauthority.org/planning/PlanningInitsIndividual

.asp?action=ViewInit&InitID=152.

Bower, Sarah, Sinikka Elliott, and Joslyn Brenton. 2014. "The Joy of Cooking." *Contexts* 13 (3): 20–23.

Brooklyn Queens Land Trust. n.d. "About BQLT." http://www.bqlt.org/About/.

Brown, Kate, and Andrew Jameton. 2000. "Public Health Implications of Urban Agriculture." *Journal of Public Health Policy* 21 (1): 20–39.

Buckley, Jennifer, and Christopher H. Peterson. n.d. "Preliminary Cost-Benefit Analysis for Urban Agriculture." Community Food Systems Toolkit, http://fyi.uwex.edu/foodsystemstoolkit/preliminary-cost-benefit-analysis/.

Canning, Patrick. 2011. *A Revised and Expanded Food Dollar Series: A Better Understanding of Our Food Costs*. Washington, DC: US Department of Agriculture Economic Research Service.

Cantrell, Patty, Kathryn Colasanti, Laura Goddeeris, Sarah Lucas, and Matt McCauley. 2013. "Food Innovation Districts: An Economic Gardening Tool." Northwest Michigan Council of Governments, http://www.nwm.org/food-innovation-districts.

Cassady, Diana, and Vidhya Mohan. 2004. "Doing Well by Doing Good? A Supermarket Shuttle Feasibility Study." *Journal of Nutrition Education and Behavior* 36 (2): 67–70.

Caton Campbell, Marcia, and Danielle A. Salus. 2003. "Community and Conservation Land Trusts as Unlikely Partners? The Case of Troy Gardens, Madison, Wisconsin." *Land Use Policy* 20 (2): 169–80.

Cedar Rapids, Iowa. 2012. "Code of Ordinances, Section 32.04." http://library.municode.com/index.aspx?clientId=16256&stateId=15&stateName=Iowa.

Center for Urban Studies. n.d. Census tract reports. Detroit, MI: Wayne State University.

Central Detroit Christian. n.d. "Peaches & Greens Produce Market." http://centraldetroitchristian.org/what-we-do/economic-development-businesses/peaches-greens/.

Change Lab Solutions. 2009. "Healthy Mobile Vending Policies: A Win-Win for Vendors and Childhood Obesity Advocates." http://www.changelabsolutions.org/sites/default/files/MobileVending_FactSht_FINAL_20130425.pdf.

Chen, Wei-ting, Megan L. Clayton, and Anne Palmer. 2015. *Community Food Security in the United States: A Survey of the Scientific Literature*. Vol. 2. Baltimore, MD: Johns Hopkins Center for a Livable Future.

Chicago Botanic Garden. 2013. "Chicago Botanic Garden's Beginning Farmers and Ranchers Development Program Establishes Four Incubator Farms in Year Two of Three-Year Program." http://www.chicagobotanic.org/pr /release/chicago_botanic_garden_establishes_four_incubator_farms.

Chicago City Council Committee on Finance. 1996. "Authorization for Execution of Intergovernmental Agreement with Chicago Park District and Forest Preserve District of Cook County for Establishment of 'NeighborSpace.'" http://www.eatbettermovemore.org/sa/policies/pdftext /ChicagoNeighborSpace.pdf.

Chicago Metropolitan Agency for Planning. n.d. "Population Forecast." http:// www.cmap.illinois.gov/data/demographics/population-forecast.

Chicago Municipal Markets Commission. 1914. *Preliminary Report to the Mayor and Aldermen of the City of Chicago*. Chicago: Municipal Markets Commission.

Chicago Policy Research Team. 2010. *Deserted? A Policy Report on Food Access in Four South Side Chicago Neighborhoods*. Chicago: University of Chicago School of Public Policy Studies.

City of Chicago. 2013. "Mayor Emanuel Launches New 'Farmers for Chicago' Network for Chicago Urban Farmers." http://www.cityofchicago.org/city /en/depts/mayor/press_room/press_releases/2013/march_2013/mayor _emanuel_launchesnewfarmersforchicagonetworkforchicagourban.html.

City of Chicago, Chicago Park District, and Forest Preserve District of Cook County. 1998. *CitySpace: An Open Space Plan for Chicago*. https://www .cityofchicago.org/city/en/depts/dcd/supp_info/cityspace_plan.html.

City of Milwaukee. 2010. *Citywide Policy Plan*. Milwaukee: Department of City Development.

City of Milwaukee Environmental Collaboration Office. 2013. *Food Systems*. ReFreshMKE, http://www.refreshmke.com/food.html.

Civil Eats. 2016. "Leaders of Color Discuss Structural Racism and White Privilege in the Food System." http://civileats.com/2016/07/15/black-lives -matter-in-the-food-movement-too/.

Clark, Jill. 2015. "From Civic Group to Advocacy Coalition: How a Food Policy Audit Became the Tool for Change." Paper presented at the annual meeting of the Midwest Political Science Association, Chicago, April.

Clynes, Melinda. 2013. "Transportation, Access to Food Hinder Older Adults Aging in Place." Mode-Shift, http://wearemodeshift.org/transportation -access-to-food-hinder-older-adults-aging-in-place.

Cobb, Tanya Denckla, and Jason Houston. 2011. *Reclaiming Our Food: How the Grassroots Food Movement Is Changing the Way We Eat*. North Adams, MA: Storey.

Coleman-Jensen, A. 2006. "Definitions of Food Security." US Department of Agriculture Economic Research Service, http://www.ers.usda.gov/topics/food-nutrition-assistance/food-security-in-the-us/definitions-of-food-security.aspx.

Collective Impact Forum. 2014. "Collective Insights on Collective Impact." http://collectiveinsights.ssireview.org.

———. 2015. "Getting Started." https://collectiveimpactforum.org/getting-started.

Committee on World Food Security. 2012. "Coming to Terms with Terminology." Food and Agriculture Organization, http://www.fao.org/docrep/meeting/026/MD776E.pdf.

Community Services Unlimited. 2016. "About Us." http://csuinc.org/about-us/.

Consultative Group on International Agricultural Research. n.d. "Food Emissions." http://ccafs.cgiar.org/bigfacts2014/#theme=food-emissions.

Cossrow, Nicole, and Bonita Falkner. 2004. "Race/Ethnic Issues in Obesity and Obesity-Related Comorbidities." *Journal of Clinical Endocrinology and Metabolism* 89 (6): 2590–94.

Coveney, John, Andrea Begley, and Danielle Gallegos. 2012. "'Savoir Fare': Are Cooking Skills a New Morality?" *Australian Journal of Adult Learning* 52 (3): 617–42.

C. R. England. 2015. "Company History." http://www.crengland.com/about-us/company-information/company-history?page=company_history.

Dahlberg, Kenneth, Kate Clancy, Robert L. Wilson, and Jan O'Donnell. 1997. "Local Food Policy Goals and Issues." In *Strategies, Policy Approaches, and Resources for Local Food System Planning and Organizing: A Resource Guide*, chap. 4. Marine St. Croix: Minnesota Food Association.

Davis, Christopher, and Biing-Hwan Lin. 2005. *Factors Affecting U.S. Beef Consumption*. Washington, DC: US Department of Agriculture, Economic Research Service.

Day-Farnsworth, Lindsey. 2016. "More Than the Sum of Their Parts: An Exploration of the Connective and Facilitative Functions of Food Policy Councils." In *Cities of Farmers: Problems, Possibilities and Processes of Producing Food in Cities*, edited by J. Dawson and A. Morales, 245–64. Iowa

City: University of Iowa Press.

Day-Farnsworth, Lindsey, Brent McCown, Michelle Miller, and Anne Pfeiffer. 2009. *Scaling Up: Meeting the Demand for Local Food*. Madison: University of Wisconsin Center for Integrated Agricultural Systems.

Day-Farnsworth, Lindsey, and A. Morales. 2011. "Satiating the Demand: Planning for Alternative Models of Regional Food Distribution." *Journal of Agriculture, Food Systems and Community Development* 2 (1): 227–47.

Day-Farnsworth, Lindsey, and D. Nelson. 2012. "Maximizing Freight Movements in Local Food Markets: An Exploration of Scale-Appropriate Solutions for Local Food Distribution." Paper presented at Wisconsin Local Food Network Summit, Delavan, WI.

Day-Farnsworth, Lindsey, Amy Bruner Zimmerman, and Jess Daniel. 2012. *Making Good Food Work Conference Proceedings*. Washington, DC: US Department of Agriculture.

Dimitri, Carolyn, Lydia Oberholtzer, and Andy Pressman. 2016. "Urban Agriculture: Connecting Producers with Consumers." *British Food Journal* 118 (3): 603–17.

Donofrio, Gregory Alexander. 2007. "Feeding the City." *Gastronomica* 7 (4): 30–41.

Dudley Street Neighbors. 2015. "Greater Boston Community Land Trust Network Launch." http://www.dsni.org/dsni-blog/2015/4/7/metro-boston -community-land-trust-network-launch.

Dundore, Lexa, and Alfonso Morales. 2015. *Evaluation of the City of Madison's "Double-Dollars" Incentive Program for the Purchase of Healthy Food from Farmers' Markets*. Madison: University of Wisconsin Department of Urban and Regional Planning.

Dunn, Carolyn, K. S. U. Jayaratne, Kristen Baughman, and Katrina Levine. 2014. "Teaching Basic Cooking Skills: Evaluation of the North Carolina Extension Cook Smart, Eat Smart Program." *Journal of Family and Consumer Sciences* 106 (1): 39–46.

Eastwood, Carolyn. 1991. *Chicago's Jewish Street Peddlers*. Chicago: Jewish Historical Society.

Ela, Nate. In press. "Urban Commons as Property Experiment: Mapping Chicago's Farms and Gardens." *Fordham Urban Law Journal*.

Eshel, Shuli, and Roger Schatz. 2004. *Jewish Maxwell Street Stories*. Charleston, SC: Arcadia.

Ewert, Brianna. 2012. "Incubating New Farmers." Master's thesis, University of Montana, Missoula.

Experimental Station. n.d. "Link Up Illinois." https://experimentalstation.org /linkup-overview.

Fair Food Network. 2013. "Strengthening Detroit Voices: Telephone Town Hall Summary." http://www.fairfoodnetwork.org/sites/default/files/FFN _SDV%20TTH%20Infographic_Print.pdf.

Farmer's Kitchen. 2014. "About." https://www.facebook.com/hollywoodfk /info?tab=page_info.

Fischer, Joan. 2012. "From Field to Food Bank." *Grow*, Spring 2012.

Fitzgerald, Kate, Lucy Evans, and Jessica Daniel. 2010. *The National Sustainable Agriculture Coalition Guide to USDA Funding for Local and Regional Food Systems*. Washington DC: National Sustainable Agriculture Coalition. http:// sustainableagriculture.net.

Fogelman, Randall. 2009. *Mo' Bucks Pilot Program Final Report*. Ann Arbor, MI: Eastern Market Corporation.

Food Chain Workers Alliance. 2012. *The Hands That Feed Us: Challenges and Opportunities for Workers along the Food Chain*. http://foodchainworkers.org /wp-content/uploads/2012/06/Hands-That-Feed-Us-Report.pdf.

Food:Land:Opportunity. 2015. "Strengthening the Resiliency of the Chicago Foodshed." http://www.cct.org/about/partnerships_initiatives/searle -foodlandopportunity/.

Food Marketing Institute. 2008. "Competition and Profit." http://www.fmi.org /docs/facts-figures/competitionandprofit.pdf?sfvrsn=2.

Franklin County Local Food Council. 2014. "The Franklin County Food Policy Audit." http://www.fclocalfoodcouncil.org/food-policy-franklin-county -resources/.

Freedgood, Julia, Marisol Pierce-Quiñonez, and Kenneth A. Meter. 2011. "Emerging Assessment Tools to Inform Food System Planning." *Journal of Agriculture, Food Systems, and Community Development* 2 (1): 83–104.

Gallagher, Mari. 2007. *Examining the Impact of Food Deserts on Public Health in Detroit*. Chicago: M. Gallagher Research & Consulting Group.

Galster, George, and Erica Raleigh. 2011. *Quantitative Indicators of Detroit Neighborhood Contexts for Child Development: Final Report*. Detroit, MI: Skillman Foundation. http://datadrivendetroit.org/wp-content /uploads/2010/04/QuantIndFINAL_042911.pdf.

Goodman, David, Erna Melanie Du Puis, and Michael K. Goodman. 2013. *Alternative Food Networks: Knowledge, Practice, Politics*. New York: Routledge.

Goodwin, Arthur E. 1929. *Markets: Public and Private, Their Establishment and Administration*. Seattle, WA: Montgomery.

Gottlieb, Robert, and Anupama Joshi. 2010. *Food Justice*. Cambridge, MA: MIT Press. http://public.eblib.com/choice/publicfullrecord.aspx?p=3339193.

Greer, Danielle, Dennis Baumgardner, Farrin Bridgewater, David Frazer, Courtenay Kessler, Erica LeCounte, Geoffrey Swain, and Ron Cisler. 2013. *Milwaukee Health Report 2012: Health Disparities in Milwaukee by Socioeconomic Status*. Milwaukee, WI: Center for Urban Population Health.

———. 2014. *Milwaukee Health Report 2013: Health Disparities in Milwaukee by Socioeconomic Status*. Milwaukee, WI: Center for Urban Population Health.

Grier, Sonya, and Shiriki Kumanyika. 2008. "The Context for Choice: Health Implications of Targeted Food and Beverage Marketing to African Americans." *American Journal of Public Health* 98 (9): 1616–29.

Growing Power. 2013. "Farmers for Chicago." http://www.growingpower.org /education/chicago-farms-and-projects/farmers-for-chicago/.

Guion, Lisa A., and Heather Kent. 2005. *Ethnic Marketing: A Strategy for Marketing Programs to Diverse Audiences*. Gainsville: University of Florida Institute of Food and Agricultural Sciences.

Gunders, Dana. 2012. "Wasted: How America Is Losing Up to 40 Percent of Its Food from Farm to Fork to Landfill." Natural Resources Defense Council, https://www.nrdc.org/food/files/wasted-food-ip.pdf.

Guthman, Julie. 2008. "Bringing Good Food to Others: Investigating the Subjects of Alternative Food Practice." *Cultural Geographies* 15: 431–47.

———. 2011. "If They Only Knew: The Unbearable Whiteness of Alternative Food." In *Cultivating Food Justice: Race, Class and Sustainability*, edited by Alison Hope Alkon and Julian Agyeman, 263–82. Cambridge, MA: MIT Press.

Hall, Kevin D., Juen Guo, Michael Dore, and Carson C. Chow. 2009. "The Progressive Increase of Food Waste in America and Its Environmental Impact." *PLOS [Public Library of Science] One* 4 (11): e7940.

Hamm, Michael W., and Anne C. Bellows. 2002. "US-Based Community Food Security: Influences, Practice, Debate." *Journal for the Study of Food and Society* 6 (1): 31–44.

Hamrick, Karen, Margaret Andrews, Joanne Guthrie, David Hopkins, and

Ket McClelland. 2011. *How Much Time Do Americans Spend on Food?* Washington, DC: US Department of Agriculture Economic Research Service.

Hanleybrown, Fay, John Kania, and Mark Kramer. 2012. "Channeling Change: Making Collective Impact Work." *Stanford Social Innovation Review*, https://ssir.org/articles/entry/channeling_change_making_collective_impact_work.

Hardesty, Shermain, Gail Feenstra, David Visher, Tracy Lerman, Dawn Thilmany-McFadden, Allison Bauman, Tom Gillpatrick, and Gretchen Nurse Rainbolt. 2014. "Values-Based Supply Chains: Supporting Regional Food and Farms." *Economic Development Quarterly* 28 (1): 17–27.

Harper, Alethea, Annie Shattuck, Eric Holt-Giménez, Alison Alkon, and Frances Lambrick. 2009. *Food Policy Councils: Lessons Learned.* Oakland, CA: Food First.

Hartman Group. 2008. *Consumer Understanding of Buying Local.* Bellevue, WA: Pulse.

Harwood, Richard C. 2015. "Putting Community in Collective Impact." Collective Impact Forum, http://collectiveimpactforum.org.

Haynes-Maslow, Lindsey. 2013. *A Qualitative Study of Perceived Barriers to Fruit and Vegetable Consumption among Low-Income Populations, North Carolina, 2011.* Atlanta, GA: Centers for Disease Control and Prevention.

Healthy Schools Campaign. n.d. "Chicago Leads the Way in Farm to School Innovation." https://healthyschoolscampaign.org/blog/chicago-leads-the-way-in-farm-to-school-innovation.

Heartland Alliance. 2012. "Heartland Human Care Services Breaks Ground on West Side Urban Farm." https://www.heartlandalliance.org/press_release/urban-farm/.

Hedden, Walter P. 1929. *How Great Cities Are Fed.* New York: Heath.

Helms, Matt. 2015. "Meijer Opens Its 2nd Detroit Store amid Song, Donations." *Detroit Free Press*, June 11. http://www.freep.com/story/news/local/michigan/detroit/2015/06/11/meijer-second-store-detroit/71062968/.

Helphand, Ben. 2015. "Permanently Grassroots with NeighborSpace." *Cities and the Environment* 8 (2): 19. http://digitalcommons.lmu.edu/cate/vol8/iss2/19.

Henry, Helen, K. Reimer, C. Smith, and M. Reicks. 2006. "Associations of Decisional Balance, Processes of Change, and Self-Efficacy with Stages of Change for Increased Fruit and Vegetable Intake among Low-Income, African-American Mothers." *Journal of the American Dietetic Association* 106

(6): 841–49.

Hodgson, Kimberly. 2012. *Planning for Food Access and Community-Based Food Systems: A National Scan and Evaluation of Local Comprehensive and Sustainability Plans*. Chicago: American Planning Association.

Hodgson, Kimberly, Marcia Caton Campbell, and Martin Bailkey. 2011. "Urban Agriculture: Growing Healthy, Sustainable Places." Chicago: American Planning Association.

Homeboy Industries. 2013. "25 Years of Widening the Circle of Compassion." http://www.homeboyindustries.org.

Hung, Yvonne. 2004. "East New York Farms: Youth Participation in Community Development and Urban Agriculture." *Children, Youth, and Environments* 14 (1): 56–85.

Institute for Agriculture and Trade Policy. 2012. "Draft Principles of Food Justice." www.iatp.org/documents/draft-principles-of-food-justice.

Iowa Department of Public Health. n.d. "Nutritional Environment Measures Survey—Vending." http://www.nems-v.com/Index.html.

James, Delores. 2004. "Factors Influencing Food Choices, Dietary Intake, and Nutrition-Related Attitudes among African Americans: Application of a Culturally Sensitive Model." *Ethnicity and Health* 9 (4): 349–67.

Jayaraman, Sarumathi, and Eric Schlosser. 2013. *Behind the Kitchen Door*. Ithaca, NY: ILR [Industrial and Labor Relations] Press.

Johns Hopkins Center for a Livable Future. 2010. "Teaching the Food System from Farm to Fork." http://foodspanlearning.org.

Kania, John, and Mark Kramer. 2011. "Collective Impact." *Stanford Social Innovation Review* 9 (1): 36–41.

———. 2013. "Embracing Emergence: How Collective Impact Addresses Complexity." *Stanford Social Innovation Review*, https://ssir.org/articles /entry/embracing_emergence_how_collective_impact_addresses_ complexity.

Karp, David. 2012. "Hollywood Farmers' Market CEO Is Fired." *Los Angeles Times*, April 6.

Khanmalek, Azeen. 2013. "CCLT Investment Strategy." *American Planning Association Louisiana Chapter Monthly Bulletin*, October.

King, Robert P., Michael S. Hand, Gigi DiGiacomo, Kate Clancy, Miguel I. Gomez, Shermain D. Hardesty, Larry Lev, and Edward W. McLaughlin. 2010. *Comparing the Structure, Size, and Performance of Local and Mainstream Food*

Supply Chains. Washington, DC: US Department of Agriculture Economic Research Service.

Kloppenburg, Jack Jr., John Hendrickson, and G. W. Stevenson. 1996. "Coming in to the Foodshed." *Agriculture and Human Values* 13 (3): 33–42.

Koc, Mustafa, Rod MacRae, Ellen Desjardins, and Wayne Roberts. 2008. "Getting Civil about Food: The Interactions between Civil Society and the State to Advance Sustainable Food Systems in Canada." *Journal of Hunger and Environmental Nutrition*, 3 (2–3): 122–44.

Korzenny, Felipe, and Betty Ann Korzenny. 2005. *Hispanic Marketing: A Cultural Perspective.* New York: Routledge.

Krome, Margaret, and George Reistad. 2014. *Building Sustainable Farms, Ranches, and Communities: A Guide to Federal Programs for Sustainable Agriculture, Forestry, Entrepreneurship, Conservation, Food Systems, and Community Development.* National Center for Appropriate Technology, https://attra.ncat.org/attra-pub/summaries/summary.php?pub=279.

Kumanyika, Shiriki Kinika, and Sonya Grier. 2006. "Targeting Interventions for Ethnic Minority and Low-Income Populations." *Future of Children* 16 (1): 187–207.

Kures, Matt. 2013. *Foundational Research for the Transform Milwaukee Initiative.* Madison: University of Wisconsin Extention.

Lawson, Laura. 2005. *City Bountiful: A Century of Community Gardening in America.* Berkeley: University of California Press.

Leib, Emily Broad. 2012a. *Good Laws, Good Food: Putting Local Food Policy to Work for Our Communities.* Jamaica Plain, MA: Harvard Law School Food Law and Policy Clinic. http://blogs.law.harvard.edu/foodpolicyinitiative/files/2011/09/FINAL-LOCAL-TOOLKIT2.pdf.

———. 2012b. *Good Laws, Good Food: Putting State Food Policy to Work for Our Communities.* Jamaica Plain, MA: Harvard Law School Food Law and Policy Clinic.

Lengnick Laura, Michelle Miller, and Gerald G. Marten. 2015. Metropolitan Foodsheds: A Resilient Solution to the Climate Change Challenge? *Journal of Environmental Studies and Sciences* 5 (4): 573–92.

Lerman, Tracy. 2012. *A Review of Scholarly Literature on Values-Based Supply Chains.* Davis, CA: University of California Agricultural Sustainability Institute.

Lev, Larry, and G. W. Stevenson. 2011. "Acting Collectively to Develop Midscale

Food Value Chains." *Journal of Agriculture, Food Systems, and Community Development* 1 (4): 119–28.

Lieb, David A. 2016. "Over 1 Million Face Loss of Food Aid over Work Requirements." Associated Press, January 30, http://onlineathens.com /national-news/2016-01-30/over-1-million-face-loss-food-aid-over-work -requirements.

Liu, Yvonne Yen, and Dominique Apollon. 2011. *The Color of Food*. Applied Research Center, http://arc.org/downloads/food_justice_021611_F.pdf.

Los Angeles Times. 2014. "Mapping L.A. Neighborhoods." http://maps.latimes .com/neighborhoods.

Lovell, S. 2010. "Multifunctional Urban Agriculture for Sustainable Land Use Planning in the United States." *Sustainability* 2 (8): 2499–2522.

Low, Sarah A., Aaron Adalja, Elizabeth Beaulieu, Nigel Key, Steve Martinez, Alex Melton, Agnes Perez, Katherine Ralston, Hayden Stewart, Shellye Suttles, Stephen Vogel, and Becca B. R. Jablonski. 2015. *Trends in U.S. Local and Regional Food Systems*. Washington, DC: US Department of Agriculture Economic Research Service.

Low, Sarah A., and Stephen Vogel. 2011. *Direct and Intermediated Marketing of Local Foods in the United States*. Washington, DC: US Department of Agriculture Economic Research Service.

Lucan, Sean C., Frances K. Barg, and Judith A. Long. 2010. "Promoters and Barriers to Fruit, Vegetable, and Fast-Food Consumption among Urban, Low-Income African Americans—A Qualitative Approach." *American Journal of Public Health* 100 (4): 631–35.

Mancino, Lisa, and Jean Kinsey. 2008. *Is Dietary Knowledge Enough? Hunger, Stress, and Other Roadblocks to Healthy Eating*. Washington, DC: US Department of Agriculture Economic Research Service.

Martin, Katie S., and Ann M. Ferris. 2007. "Food Insecurity and Gender Are Risk Factors for Obesity." *Journal of Nutrition Education and Behavior* 39 (1): 31–36.

Martinez, Steve, Michael Hand, Michelle Da Pra, Susan Pollack, Katherine Ralston, Travis Smith, Stephen Vogel, Shellye Clark, Luanne Lohr, Sarah Low, and Constance Newman. 2010. *Local Food Systems: Concepts, Impacts, and Issues*. Washington, DC: US Department of Agriculture Economic Research Service.

Matsunaga, Michael. 2008. *Concentrated Poverty in Los Angeles*. Los Angeles: Economic Roundtable.

McCabe, Margaret Sova. 2011. "Foodshed Foundations: Law's Role in Shaping Our Food System's Future." *Fordham Environmental Law Review* 22: 563.

McCormack, Lacey Arneson, Melissa Nelson Laska, Nicole I. Larson, and Mary Story. 2010. "Review of the Nutritional Implications of Farmers' Markets and Community Gardens: A Call for Evaluation and Research Efforts." *Journal of American Dietetic Association*, 110 (3): 399–408.

McGowan, Joshua. 2011. "Communities Aren't 'Food Deserts,' but Healthy Eating Eludes Many." Milwaukee Neighborhood News Service, http://milwaukeenns.org/2011/09/06/communities-arent-food-deserts-but-healthy-eating-eludes-many-3/.

McKinnon, Robin, Jill Reedy, Meredith Morrissette, Leslie Lytle, and Amy Yaroch. 2009. "Measures of the Food Environment: A Compilation of the Literature, 1990–2007." *American Journal of Preventive Medicine* 36 (4): 124–33.

McMillan, Tracie. 2014. "Can Whole Foods Change the Way Poor People Eat?" *Slate*, November 19. http://www.slate.com/articles/life/food/2014/11/whole_foods_detroit_can_a_grocery_store_really_fight_elitism_racism_and.html.

Meter, Ken. 2009. "Mapping the Minnesota Food Industry." Minneapolis, MN: Crossroads Resource Center.

Metro Regional Transit Authority. n.d. "Grocery Shopping Bus 91-95." Akron, OH: Akron Metro.

Meyers, R. 2015. *Residential Organics Collection Feasibility Report*. Milwaukee, WI: Department of Public Works.

Millennium Ecosystem Assessment. 2005. *Ecosystems and Human Well-Being: Synthesis*. Washington, DC: Island Press.

Miller, Pepper, and Herb Kemp. 2006. *What's Black About It? Insights to Increase Your Share of a Changing African-American Market*. Ithaca, NY: Paramount Market.

Miller, Richard K., and Kelli Washington, eds. 2014. "Buying Local." In *Consumer Behavior*, 108–11. Loganville, GA: Richard K. Miller & Associates.

Mohan, Vidhya, and Diana Cassady. 2002. *Supermarket Shuttle Programs: A Feasibility Study for Supermarkets Located in Low-Income, Transit-Dependent, Urban Neighborhoods in California*. Davis: University of California Center for Advanced Studies in Nutrition and Social Marketing.

Moore, Latetia, and Frances Thompson. 2013. *Adults Meeting Fruit and Vegetable*

Intake Recommendations—United States, 2013. Atlanta, GA: Centers for Disease Control and Prevention.

Morales, Alfonso. 1993. *Making Money at the Market: The Social and Economic Logic of Informal Markets.* Evanston, IL: Northwestern University Department of Sociology.

———. 2000. "Peddling Policy: Street Vending in Historical and Contemporary Context." *International Journal of Sociology and Social Policy* 20 (3/4): 14.

———. 2002. "Radio Mercado: Station Format and Alternative Models of the Audience in the U.S.-Mexico Border Region." *Journal of Borderlands Studies* 17 (1): 79–102.

———. 2009a. "Public Markets as Community Development Tools." *Journal of Planning Education and Research* 28 (4): 426–40.

———. 2009b. "A Woman's Place Is on the Street: Purposes and Problems of Mexican American Women Entrepreneurs." In *Wealth Creation and Business Formation among Mexican-Americans: History, Circumstances and Prospects,* edited by John S. Butler, Alfonso Morales, and David Torres, 99–125. West Lafayette, IN: Purdue University Press.

———. 2011a. "Growing Food *and* Justice: Dismantling Racism through Sustainable Food Systems." In *Cultivating Food Justice: Race, Class and Sustainability,* edited by Alison Hope Alkon and Julian Agyeman, 149–76. Cambridge, MA: MIT Press.

———. 2011b. "Public Markets: Prospects for Social, Economic, and Political Development." *Journal of Planning Literature* 26 (3): 3–17.

———. 2012. "Understanding and Interpreting Tax Compliance Strategies among Street Vendors." In *The Ethics of Tax Evasion: Perspectives in Theory and Practice,* edited by Robert McGee, 83–106. Dordrecht, Netherlands: Springer.

Morales, Alfonso, Steve Balkin, and Joe Persky. 1995. "Contradictions and Irony in Policy Research on the Informal Economy: A Reply." *Economic Development Quarterly* 9 (4): 327–30.

Morales, Alfonso, and Gregg Kettles. 2009a. "Healthy Food Outside: Farmers' Markets, Taco Trucks, and Sidewalk Fruit Vendors." *Journal of Contemporary Health Law and Policy* 26: 20.

———. 2009b. "Zoning for Markets and Street Merchants." *Zoning Practice* 25 (1): 1–8.

———. 2009c. "Zoning for Public Markets and Street Vendors." *Zoning*

Practice 25 (2): 1–8.

Mukherji, Nina, and Alfonso Morales. 2010. "Zoning for Urban Agriculture."
Zoning Practice 26 (3): 1–8.

Muller, Mark, Angie Tagtow, Susan L. Roberts, and Erin MacDougall. 2009.
"Aligning Food Systems Policies to Advance Public Health." *Journal of
Hunger and Environmental Nutrition* 4 (3–4): 225–40.

Natchez, Meryl. 2011. "The Amazing Original Homemade Compost Buster."
Dactyls & Drakes, http://www.dactyls-and-drakes.com/?s=compost+buster.

National Grocers Association. 2012. *National Grocers Association 2012
Supermarket Guru Consumer Panel Survey.* Arlington, VA: National Grocers
Association.

Neuner, Kailee, Sylvia Kelly, and Samina Raja. 2011. *Planning to Eat? Innovative
Local Government Plans and Policies to Build Healthy Food Systems in the
United States.* Buffalo: State University of New York.

New State Ice Co. v. Liebmann. 1932. 285 U.S. 262.

New York Restoration Project. n.d. "Target East Harlem Community Garden."
https://www.nyrp.org/green-spaces/garden-details/target-east-harlem
-community-garden.

North Central Cooperative Extension Association. 2011. *Report of the North
Central Cooperative Extension Association Metropolitan Food Systems
Symposium: Draft Executive Summary.* http://www.nccea.org.

O'Brien, Jennifer, and Tanya Denckla Cobb. 2012. "The Food Policy Audit: A
New Tool for Community Food System Planning." *Journal of Agriculture,
Food Systems, and Community Development* 2 (3): 177–91. http://dx.doi
.org/10.5304/jafscd.2012.023.002.

Opoien, Jessie. 2014. "Wisconsin One of Three States to Reject Food-Stamp
Increase." *Capital Times.* September 18. http://host.madison.com/news
/local/writers/jessie-opoien/wisconsin-one-of-three-states-to-reject-food
-stamp-increase/article_ac1cdac8-ef38-55cf-bac6-732341209772.html.

Organic Authority. 2012. "Earth Day Profile: Chef Ernest Miller on a Different
Kind of Soul Food." April 25. http://www.organicauthority.com/restaurant
-buzz/earth-day-ernest-miller-the-farmers-kitchen.html.

Otto, Jayson, ed. 2016. "Municipal Housekeepers and the High Cost of Living:
The Establishment of Gardening Programs and Farmers' Markets by Grand
Rapids Women's Clubs in the Early Twentieth Century." In *Cities of Farmers:
Problems, Possibilities and Processes of Producing Food in Cities,* 21–38. Iowa

City: University of Iowa Press.

Packer, Melina. 2014. "Civil Subversion: Making 'Quiet Revolution' with the Rhode Island Food Policy Council." *Journal of Critical Thought and Praxis* 3 (1): 6. http://lib.dr.iastate.edu/jctp/vol3/iss1/6.

Patel, Raj. 2008. *Stuffed and Starved: The Hidden Battle for the World Food System.* New York: Melville House.

Patton, Michael Quinn. 2011. *Developmental Evaluation: Applying Complexity Concepts to Enhance Innovation and Use.* New York: Guilford Press.

Paxton, A. 1994. *The Food Miles Report: The Dangers of Long-Distance Food Transport.* London: Safe Alliance. http://www.sustainweb.org /publications/?id=191.

PD&R Edge. 2012. "Community Land Trusts in Atlanta, Georgia: A Central Server Model." http://www.huduser.gov/portal/pdredge/pdr_edge _inpractice_112312.html.

Peters, Christian, Nelson Bills, Arthur Lembo, Jennifer Wilkins, and Gary Fick. 2009. "Mapping Potential Foodsheds in New York State: A Spatial Model for Evaluating the Capacity to Localize Food Production." *Renewable Agriculture and Food Systems* 24 (1): 72–84.

———. 2012. "Mapping Potential Foodsheds in New York State by Food Group: An Approach for Prioritizing Which Foods to Grow Locally." *Renewable Agriculture and Food Systems* 27 (2): 125–37.

Phills, James A. Jr., Kriss Deiglmeier, and Dale T. Miller. 2008. "Rediscovering Social Innovation." *Stanford Social Innovation Review* 6 (4): 34–43.

Platt, Brenda, Bobby Bell, and Cameron Harsh. 2013. *Pay Dirt: Composting in Maryland to Reduce Waste, Create Jobs, and Protect the Bay.* Institute for Local Self-Reliance, http://www.ilsr.org/wp-content/uploads/2013/05/ILSR-Pay -Dirt-Report-05-11-13.pdf.

PolicyLink and Local Initiatives Support Corporation. 2008. *Grocery Store Attraction Strategies: A Resource Guide for Community Activists and Local Governments.* http://community-wealth.org/sites/clone.community-wealth .org/files/downloads/tool-policylink-lisc-grocery.pdf.

PolicyLink, Reinvestment Fund, and Food Trust. n.d. "Making the Case: Why Healthy Food Access Matters." Healthy Food Access Portal, http://www .healthyfoodaccess.org/get-started/making-the-case.

Pollan, Michael. 2008. *In Defense of Food: An Eater's Manifesto.* New York: Penguin Press.

Pothukuchi, Kami. 2011. *The Detroit Food System Report, 2009–2010*. Detroit, MI: Wayne State University.

Pothukuchi, Kami, and Jerome L. Kaufman. 1999. "Placing the Food System on the Urban Agenda: The Role of Municipal Institutions in Food Systems Planning." *Agriculture and Human Values* 16 (2): 213–24.

———. 2000. "The Food System: A Stranger to the Planning Field." *Journal of the American Planning Association* 66 (2): 113–24.

Pullman, Madeleine, and Zhaohui Wu. 2012. *Food Supply Chain Management: Economic, Social and Environmental Perspectives*. New York: Routledge.

Raja, Samina, Branden Born, and Jessica Russell. 2008. *A Planners' Guide to Community and Regional Food Planning: Transforming Food Environments, Facilitating Healthy Eating*. Chicago: American Planning Association.

Raja, Samina, Changxing Ma, and Pavan Yadav. 2008. "Beyond Food Deserts: Measuring and Mapping Racial Disparities in Neighborhood Food Environments." *Journal of Planning Education and Research* 27 (4): 469–82.

Ramde, Dinesh. 2011. "Food Pantries Request Healthier Donations over Bulk Junk Food This Christmas." *Huffington Post*, November 21.

Rhode Island Department of Environmental Management. n.d. *Forest Stewardship: Rhode Island Landowners Discover New Strategies in Forest Conservation*. http://www.dem.ri.gov/programs/bnatres/forest/pdf/forstew .pdf.

Ringstrom, Eva, and Branden Born. 2011. *Food Access Policy and Planning Guide*. UW [University of Washington] Northwest Center for Livable Communities, http://www.nyc.gov/html/ddc/downloads/pdf/ActiveDesignWebinar /King%20County%20Food%20Access%20Guide.pdf.

Roberts, Wayne. 2008. *The No-Nonsense Guide to World Food*. Oxford, UK: New Internationalist Publications.

Rosenberg, Greg. 2007. *Troy Gardens Case Study*. Madison Area Community Land Trust, http://www.greenfordable.com/troygardens/.

Roubal, Anne, and Alfonso Morales, eds. 2016. "Chicago Marketplaces: Advancing Access to Healthy Food." In *Cities of Farmers: Problems, Possibilities and Processes of Producing Food in Cities*, 191–211. Iowa City: University of Iowa Press.

Sauer, Andrea. 2012. "FPC [Food Policy Council] List Update Analysis." Community Food Security Coalition, http://www.foodsecurity.org.

Schaeffer, Leanne, and Brian Miller. 2012. "Should Future Dietetic Graduates

Know How to Cook?" *Journal of Foodservice Management and Education* 6 (1): 25–30.

Schiff, Rebecca. 2008. "The Role of Food Policy Councils in Developing Sustainable Food Systems." *Journal of Hunger and Environmental Nutrition* 3 (2–3): 206–28.

Schmitz, Paul. 2014. "The Culture of Collective Impact." *Huffington Post*, October 22.

Schneggenburger, Andy. 2011. Bringing CLTs [Community Land Trusts] to Scale in Atlanta. *Shelterforce*, February 7. http://www.shelterforce.org/article/bringing_clts_to_scale_in_atlanta/.

Sears, Michael. 2011. "Poverty Numbers Spike in Milwaukee." *Milwaukee Journal Sentinel*, September 21.

Shannon, Alan. 2014. "From Small Potatoes to 36,000 Pounds of Carrots: Farm to School Grows." US Department of Agriculture, http://blogs.usda.gov/2014/01/27/from-small-potatoes-to-36000-pounds-of-carrots-farm-to-school-grows/.

Shapiro, Ilana. 2002. "Training for Racial Equity and Inclusion: A Guide to Selected Programs." Alliance for Conflict Transformation, http://www.racialequitytools.org/resourcefiles/shapiro.pdf.

Sherman, Caroline B. 1937. "Markets, Municipal." In *Encyclopedia of the Social Sciences*, edited by Alvin S. Johnson and Edward R. A. Seligman, 139–44. New York: Macmillan.

Siedenburg, Kai, and Kami Pothukuchi, eds. 2002. "What's Cooking in Your Food System? A Guide to Community Food Assessment." Community Food Security Coalition, http://foodsecurity.org/whats_cooking/.

Skeo Solutions. 2012. *Urban Agriculture Code Audit, Milwaukee, Wisconsin*. Washington, DC: U.S. Environmental Protection Agency.

Skid, Nathan. 2011. "Whole Foods Market Browses in Midtown." Crain's Detroit Business, http://www.crainsdetroit.com/article/20110403/FREE/304039993/whole-foods-market-browses-in-midtown.

Slocum, Rachel. 2006. "Antiracist Practice and the Work of Community Food Organizations." *Antipode* 38: 327–49.

Small Business Administration. 2012. "Frequently Asked Questions." http://www.sba.gov/sites/default/files/FAQ_Sept_2012.pdf.

Snowden, Mary. 2006. "Farm Profile: Pak Express Farm." Farm Fresh, http://www.farmfresh.org/food/farm.php?farm=766#profile.

Social Justice Learning Institute. n.d. "Educational Equity." http://sjli.org
/program/educational-equity.

Southside Community Land Trust. n.d. "Urban Edge Farm." http://www
.southsideclt.org/urbanedge.

Spitzer, Theodore Morrow, and Hilary Baum. 1995. *Public Markets and
Community Revitalization*. Washington, DC: Urban Land Institute.

Stevenson, G.W., and Rich Pirog. 2008. "Values-Based Food Supply Chains:
Strategies for Agri-food Enterprises of the Middle." In *Food and the Mid-Level
Farm: Renewing an Agriculture of the Middle*, edited by Thomas A. Lyson, G.
W. Stevenson, and Rick Welsh, 119–45. Cambridge, MA: MIT Press.

Suerth, Lauren, and Alfonso Morales. 2014. "Zoning for Small-Scale
Composting in Urban Areas." *Zoning Practice* 31 (9): 1–8.

Sugrue, Thomas J. 2005. *The Origins of the Urban Crisis: Race and Inequality in
Postwar Detroit*. Princeton, NJ: Princeton University Press. First published in
1996.

Sullivan, J. W. 1913. *Markets for the People: The Consumer's Part*. New York:
Macmillan.

Sweet, M. L. 1961. "History of Municipal Markets: Has Lessons for Commercial
Renewal Projects." *Journal of Housing* 18: 237–47.

Tangires, Helen. 2003. *Public Markets and Civic Culture in Nineteenth-Century
America*. Baltimore, MD: Johns Hopkins University Press.

Taylor, John R., and Sarah Taylor Lovell. 2012. "Mapping Public and Private
Spaces of Urban Agriculture in Chicago through the Analysis of High-
Resolution Aerial Images in Google Earth." *Landscape and Urban Planning*
108: 57–70.

Teig, Ellen, Joy Amulya, Lisa Bardwell, Michael Buchenau, Julie Marshall,
and Jill Litt. 2009. "Collective Efficacy in Denver, Colorado: Strengthening
Neighborhoods and Health through Community Gardens." *Health and Place*
15 (4): 1115–22.

Tharp, Marye C. 2001. *Marketing and Consumer Identity in Multiracial America*.
Thousand Oaks, CA: Sage.

Thomas, Chris D., Alison Cameron, Rhys E. Green, Michel Bakkenes, Linda
J. Beaumont, Yvonne C. Collingham, Barend F. Erasmus, Marinez Ferreira
de Siqueira, Alan Grainger, Lee Hannah, Lesley Hughes, Brian Huntley,
Albert S. van Jaarsveld, Guy F. Midgley, Lera Miles, Miguel A. Ortega-Huerta,
A. Townsend Peterson, Oliver L. Phillips, and Stephen E. Williams. 2004.

"Extinction Risk from Climate Change." *Nature* 427 (6970): 145–48.

Thomas, Richard W. 1992. *Life for Us Is What We Make It: Building Black Community in Detroit, 1915–1945*. Bloomington: Indiana University Press.

Thompson, D. 2013. "Why Economics Is Really Called 'the Dismal Science.'" *Atlantic*, December 17. http://www.theatlantic.com /business/archive/2013/12/why-economics-is-really-called-the-dismal -science/282454/.

Treiman, Katherine, Vicki Freimuth, Dorothy Damron, Anita Lasswell, Jean Anliker, Stephen Havas, Patricia Langenberg, and Robert Feldman. 1996. "Attitudes and Behaviors Related to Fruits and Vegetables among Low-income Women in the WIC Program." *Journal of Nutrition Education and Behavior* 28 (3): 149–56.

Treuhaft, Sarah, and Allison Karpyn. 2010. *The Grocery Gap: Who Has Access to Healthy Food and Why It Matters*. PolicyLink and the Food Trust, http:// thefoodtrust.org/uploads/media_items/grocerygap.original.pdf.

Trust for Public Land. n.d. "Working Lands." http://www.tpl.org/working -lands-0.

Tursini, Andrea. 2010. "Farm Incubation" outline. Northeast Beginning Farmers Project, http://www.nebeginningfarmers.org/files/2012/05 /Andrea-Tursini-Farm-Incubators-sc7cbp.pdf.

University of Wisconsin Extension. 2017. *Urban Agriculture Manual*. http:// urbanagriculture.horticulture.wisc.edu.

———. n.d. *Community Food Systems Toolkit*. http://fyi.uwex.edu /foodsystemstoolkit/.

Urban Farm Pathways Project. 2015. *Englewood Community Farms Prospectus and Business Plan*. http://www.foodlandopportunity.org/downloads /Englewood_Prospectus_Business-Plan.pdf.

US Bureau of the Census. 2011. *American Community Survey*. Washington, DC: US Government Printing Office.

US Department of Agiculture. 2016. *Urban Agriculture Tool Kit*. Washington, DC: US Government Printing Office.

US Department of Agiculture Economic Research Service. 2012. "Key Statistics and Graphics." http://www.ers.usda.gov/topics/food-nutrition-assistance /food-security-in-the-us/key-statistics-graphics.aspx#foodsecure.

———. 2013. *Food Security Status of U.S. Households in 2013*. Washington, DC: US Government Printing Office.

———. n.d. "Food Expenditures." http://www.ers.usda.gov/data-products
/food-expenditures.aspx#26636.

———. n.d. "Retailing and Wholesaling." http://ers.usda.gov/topics/food
-markets-prices/retailing-wholesaling.aspx.

US Department of Agriculture National Agricultural Statistics Service. 2015.
"Quick Stats." http://quickstats.nass.usda.gov/results/58B27A06-F574-315B
-A854-9BF568F17652#7878272B-A9F3-3BC2-960D-5F03B7DF4826.

US Department of Health and Human Services and US Department of
Agriculture. 2015–2020 Dietary Guidelines for Americans. 8th edition.
December 2015. Available at http://health.gov/dietaryguidelines/2015
/guidelines.

Vallianatos, Mark, Amanda Shaffer, and Robert Gottlieb. 2002. Transportation
and Food: The Importance of Access. Los Angeles: Urban and Environmental
Policy Institute. http://departments.oxy.edu/uepi/cfj/publications
/transportation_and_food.pdf.

Ver Ploeg, Michele, Vince Breneman, Paula Dutko, Ryan Williams, Samantha
Snyder, Chris Dicken, and Phil Kaufman. 2012. "Access to Affordable and
Nutritious Food: Updated Estimates of Distance to Supermarkets Using
2010 Data." Washington, DC: US Department of Agriculture Economic
Research Service.

Vidgen, Helen A., and Danielle Gallegos. 2011. What Is Food Literacy and Does
It Influence What We Eat: A Study of Australian Food Experts. Brisbane,
Australia: Queensland University of Technology. http://eprints.qut.edu
.au/45902/.

Viscelli, Steve. n.d. "Getting It from Here to There: Urban Truck Ports and
the Coming Freight Crisis." Center on Wisconsin Strategy, unpublished
manuscript.

White, Monica M. 2010. "D-Town: African American Farmers, Food Security
and Detroit." Black Agenda Report, http://www.blackagendareport.com
/content/d-town-african-american-farmers-food-security-and-detroit.

———. 2011. "Sisters of the Soil: Urban Gardening as Resistance in Detroit."
Race/Ethnicity 5 (1): 13–28.

Wilde, Parke. 2013. Food Policy in the United States: An Introduction. New York:
Routledge.

Winne, Mark. 2008. Closing the Food Gap: Resetting the Table in the Land of
Plenty. Boston: Beacon Press.

Wisconsin Department of Agriculture, Trade, and Consumer Protection. 2014. *FY14—Farm Bill Wisconsin Specialty Crop Block Grant Program Request for Proposals and Grant Manual.* Madison: State of Wisconsin.

Worsley, Anthony, Wei Wang, Sinem Ismail, and Stacey Ridley. 2014. "Consumers' Interest in Learning about Cooking: The Influence of Age, Gender, and Education." *International Journal of Consumer Studies* 38: 258–64.

Xiao, Canming. 2014. *Focus on "Biobased," "Biodegradable," and "Compostable" Plastics.* Lacey: Washington State Department of Ecology.

Yuen, Jeffrey. 2012. "Hybrid Vigor: An Analysis of Land Tenure Arrangements in Addressing Land Security for Urban Community Gardens." Master's thesis, Columbia University, New York.

Zenk, Shannon N., Amy J. Schulz, Teretha Hollis-Neely, Richard T. Campbell, Nellie Holmes, Gloria Watkins, Robin Nwankwo, and Angela Odoms-Young. 2006. "Fruit and Vegetable Intake in African Americans: Income and Store Characteristics." *American Journal of Preventive Medicine* 29 (1): 1–9.

WEBSITES

Growing Food and Justice for All Initiative. http://growingfoodandjustice.org.

Kompost Kids. http://kompostkids.org.

People's Institute for Survival and Beyond. http://www.pisab.org.

Second Harvest Foodbank of Southern Wisconsin. http://www.secondharvestmadison.org.

University of Wisconsin Extension, Center for Community and Economic Development. http://cced.ces.uwex.edu/.

Uprooting Racism, Planting Justice. https://www.facebook.com/URPJDetroit/.

Contributors

Angie Allen, UWEX, Milwaukee County

Erika Allen, Growing Power, Chicago, and Chicago Food Policy Action Council

Martin Bailkey, CRFS project comanager and food systems planner, formerly
    with Growing Power

Stephanie Calloway, community organizer, CORE/El Centro, Milwaukee

Marcia Caton Campbell, Center for Resilient Cities

Rodger Cooley, Chicago Food Policy Action Council

Lindsey Day-Farnsworth, PhD student in environment and resources,
    University of Wisconsin–Madison

Nate Ela, PhD student in sociology, University of Wisconsin–Madison

Nicodemus Ford, University of Wisconsin Cooperative Extension

Jason Grimm, food system planner, Iowa Valley Resource Conservation &
    Development

April Harrington, Growing Home, Chicago

Margaret Krome, policy analyst, Michael Fields Agricultural Institute

Greg Lawless, CRFS project comanager, food system specialist, UWEX

Jeffrey Lewis, UWEX

Oona Mackesey-Green, CRFS communications intern, UW–Madison

Colleen McKinney, Center for Good Food Purchasing

Michelle Miller, Center for Integrated Agricultural Systems, UW–Madison

Alfonso Morales, professor of urban and regional planning, UW–Madison

Anne Pfeiffer, urban farming specialist, University of Minnesota

Samuel Pratsch, University of Wisconsin Cooperative Extension

Harry Rhodes, Growing Home, Chicago

Greg Rosenberg, principal, Rosenberg and Associates, Madison

Neelam Sharma, Community Services Unlimited

Rebekah Silverman, Growing Home, Chicago

Laurell Sims, Growing Power, Chicago, and Chicago Food Policy Action Council

Desiré Smith, former CRFS project intern, PEOPLE's urban agriculture program, UW–Madison

Shelly Strom, Community GroundWorks

Monica Theis, senior lecturer, Department of Food Science, UW–Madison

Steve Ventura, CRFS project codirector, UW–Madison

Malik Yakini, Detroit Black Community Food Security Network

Index

US legislation and regulation, differences between, 204–206

value-added crops, 58–59
values-based food supply chains, 67–70
vermicompost, 56, 141, 143
vertical farming, 196, 240
Village Market Place, Los Angeles, California, 82–84

Walnut Way Conservation Corporation, Milwaukee, Wisconsin, 244
waste, food, 97, 99, 127, 141–156, 211, 248

water, 43, 58, 152, 246
white privilege, 163–164, 185, 229, 231, 234–237
Wisconsin–Madison, University of, food programs, 1, 73, 135–140, 150, 190–199, 223
Women, Infants, and Children (WIC) program, 89, 202, 207

yields, of crops, 57, 59–61, 90, 132–133, 240

zoning, 208–210